Andrzej Marek Żak

Transmission Electron Microscopy

Also of interest

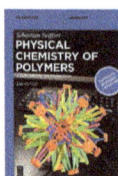

Physical Chemistry of Polymers.
A Conceptual Introduction
2nd Edition
Sebastian Seiffert, 2023
ISBN 978-3-11-071327-5, e-ISBN (PDF) 978-3-11-071326-8,
e-ISBN (EPUB) 978-3-11-071339-8

Surface Characterization Techniques.
From Theory to Research
Rawesh Kumar, 2022
ISBN 978-3-11-065599-5, e-ISBN (PDF) 978-3-11-065648-0,
e-ISBN (EPUB) 978-3-11-065658-9

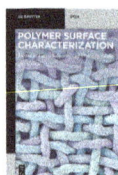

Polymer Surface Characterization.
2nd Edition
Luigia Sabbatini, Elvira De Giglio (Eds.), 2022
ISBN 978-3-11-070104-3, e-ISBN (PDF) 978-3-11-070109-8,
e-ISBN (EPUB) 978-3-11-070114-2

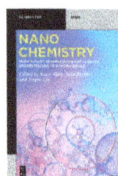

Nanochemistry.
From Theory to Application for In-Depth Understanding of Nanomaterials
Xuan Wang, Sajid Bashir, Jingbo Liu (Eds.), 2022
ISBN 978-3-11-073985-5, e-ISBN (PDF) 978-3-11-073987-9,
e-ISBN (EPUB) 978-3-11-073999-2

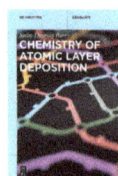

Chemistry of Atomic Layer Deposition
Seán Thomas Barry, 2021
ISBN 978-3-11-071251-3, e-ISBN (PDF) 978-3-11-071253-7,
e-ISBN (EPUB) 978-3-11-071259-9

Andrzej Marek Żak

Transmission Electron Microscopy

A Practical Guide to Using a Microscope

DE GRUYTER

Author
Professor Andrzej Marek Żak
Wroclaw University of Science and Technology (WUST)
50-370 Wroclaw
Poland
andrzej.zak@pwr.edu.pl

ISBN 978-3-11-131649-9
e-ISBN (PDF) 978-3-11-131701-4
e-ISBN (EPUB) 978-3-11-131715-1

Library of Congress Control Number: 2024940934

Bibliographic information published by the Deutsche Nationalbibliothek
The Deutsche Nationalbibliothek lists this publication in the Deutsche Nationalbibliografie;
detailed bibliographic data are available on the Internet at http://dnb.dnb.de.

www.degruyter.com

The book would not have been possible without the support of many people, I would like to mention at least a few of them.

Thank you:

    my wife Anna and son Kazimierz, for love and understanding;

    my collegues: Dominika Benkowska-Biernacka, Olga Kaczmarczyk, and Aleksandra Królicka, for the review and valuable comments;

    many fantastic colleagues that I have met along the way, for our joint journey into the world of micro and nano pieces of the matter.

Contents

1 Introduction

Transmission electron microscopy has been available to researchers interested in structural research in the fields of physics, materials science, and life sciences for over 80 years. It was the first method in history to allow the observation of objects beyond the capabilities of light microscopy. Today, after almost a century of development, the ability to observe matter at the atomic scale is no longer surprising. The benefits of using high-resolution electron observations extend to every branch of modern science.

Unfortunately, due to the relatively high complexity of the research equipment, the multifaceted nature of the interactions of the electron beam with matter, and the complicated work methodology, which requires relatively extensive knowledge about the device and method – transmission electron microscopy remains a relatively elite method in many research centers. And the use of microscopes is sometimes limited to a narrow group of well-trained operators and specialists. This is because the minimum entry threshold for independent work is set relatively high, especially in the case of slightly older microscopes. There are several excellent books and manuals, but entry-level researchers could be discouraged by the complex physical and mathematical descriptions of the phenomena occurring inside an electron microscope. Device manuals have also evolved. Back in the 1990s, you could read about the physical basis of imaging, as well as details of the device's construction and functionality, and precisely follow work protocols. Nowadays, you increasingly come across much shorter manuals, mostly focused on the effective use of microscope control software. There is nothing wrong with this; it is a normal part of popularizing the method and lowering the entry threshold to independent work. In such a case, I hope that the following book will shed some light on the theoretical, structural, and functional basis of transmission electron microscope.

As a student, PhD student, independent researcher, and laboratory head, I have repeatedly introduced new people to the world of electron microscopy techniques. This is usually accompanied by a supporting lecture followed by a relatively long practical session. A pretrained researcher usually still needs assistance over the following days and weeks to learn to deal with everyday challenges independently. Experience shows that many other laboratories operate in a very similar way, and the basis for transferring knowledge is the master-student relationship. The purpose of this book is to provide basic education to interested readers before they receive practical operator training. I hope it will also serve as one of the daily teaching aids. During my career, I have worked on over a dozen different TEM models, manufactured over a period of more than 50 years. I have tried to create a protocol that would work for each of these devices. However, further development of technology may bring innovations that will not be discussed in this book. Likewise, I have tried not to discuss in too much detail procedures that are unique to specific manufacturers

https://doi.org/10.1515/9783111317014-001

or rarely used. I hope that in such cases, each user will be able to count on the support of their microscope supervisor.

In the next Chapter 2, I will briefly describe the history of the development of electron microscopy. In Chapter 3, I will briefly discuss the structure of a TEM by analogy to a light microscope. Chapter 4 will focus on specific parts of the microscope, such as the electron gun, vacuum system, lenses, and detectors. I am convinced that basic knowledge about these components significantly facilitates subsequent work and everyday troubleshooting. In Chapter 5, I will briefly discuss various sample preparation methods, both for nanomaterials and solid samples. This will not be an exhaustive treatment of the topic, but rather a discussion of the basic techniques, their possibilities, and limitations. Chapter 6 provides a complete protocol for working with the TEM, from startup, through sample installation, electron emission, imaging, microphotograph acquisition, and finishing of work. There you will also find graphics describing the centering and calibration of various microscope systems. In Chapter 7, I discuss in more detail the topic of selected area electron diffraction (SAED) and imaging using bright field (BF) and dark field (DF) techniques. This method, using the unique features of the TEM structure, allows for the detection of the occurrence of individual crystal phases and their orientation, without resorting to atomic resolution and elemental analyses. In Chapter 8, I will discuss the most common problems and how to fix them. Chapter 9, in turn, indicates possibilities for further development and literature sources for the most interested readers. Most chapters end with synthetic take-home messages to help remember the most important information

I hope that reading this book will be the beginning of an interesting adventure for you, as well as significant help in your daily and independent work with TEM.

2 A brief history of the origins of transmission electron microscopy

The early twentieth century was a period of intensive research on the atomic structure of the matter, significantly impacting the development of microscopy. One of the pivotal discoveries was quantum theory, initiated by the work of Albert Einstein, Max Planck, and Niels Bohr. This theory suggested that light could exhibit both wave-like and particle-like properties, which was crucial for understanding the nature of matter at the subatomic level.

In 1924, Louis de Broglie introduced the concept of matter waves, proposing that particles such as electrons could exhibit wave-like behavior [1]. This French physicist, working in Paris, was awarded the Nobel Prize in 1929 for his research. His discovery was fundamental to the development of electron microscopy, as it paved the way for using electron waves to image microscopic structures.

In 1927, German physicists Clinton Davisson and Lester Germer confirmed the wave nature of electrons by conducting an electron diffraction experiment on crystals [2]. The researchers accidentally positioned a nickel sample in the path of low-energy electrons (accelerated by a voltage of approximately 50 V, a value you will soon compare with those typical for modern electron microscopes). This nickel sample, heated to consist of large single crystals, caused the electrons to diffract in accordance with Bragg's hypothesis (which describes wave diffraction on a crystal lattice). This discovery was crucial for the development of electron microscopy, as it demonstrated that electrons could interact with the crystal lattice of the material under investigation.

Around the same time, George Paget Thomson (the son of Joseph J. Thomson, Nobel Laureate in Physics in 1906) made a similar discovery. He observed the polycrystalline diffraction of high-energy electrons on thin metal films [3]. The polycrystalline diffraction patterns he obtained were correlated with those produced by X-ray methods and Bragg's diffraction hypothesis. Thomson's experiment, alongside the work of Davisson and Germer, confirmed the wave nature of electrons, representing another milestone in the development of electron microscopy. These experiments demonstrated that de Broglie waves are not merely theoretical constructs but can also be observed and utilized in practice. As a result of the work by both groups, the Nobel Prize in Physics in 1937 was awarded to Davisson and Thomson. Additionally, the experiment by Davisson and Germer was included in the twenty-first-century list of the ten most beautiful experiments in physics – *The Prism and the Pendulum: The Ten Most Beautiful Experiments in Science* by Robert P. Crease. It was the most recent experiment highlighted in this compilation.

In 1931, German physicists Ernst Ruska and Max Knoll constructed the first electron microscope, which had a magnification of approximately 7× [4]. Just two years later, Ruska refined the design, surpassing the resolving power of the light microscope.

https://doi.org/10.1515/9783111317014-002

For this achievement, he was awarded the Nobel Prize in Physics in 1986. His lecture on this occasion beautifully summarizes the early history of electron microscopy [5].

However, it is important to acknowledge other significant contributors. In 1941, the Prussian Academy of Sciences awarded the Leibniz Medal for the development of electron microscopy to seven individuals: Manfred von Ardenne, Hans Boersch, Bodo von Borries, Ernst Brüche, Max Knoll, Hans Mahl, and Ernst Ruska. This list more accurately represents the key figures who laid the foundation for this new method.

If you are interested in a comprehensive list of the most important publications in the history of electron microscopy, I highly recommend the extensive work on the subject [6], which summarizes in tabular form all the most significant historical sources. For a more narrative description, I suggest Chapter IV of [7], [8], or [9]. However, consider reading these only after familiarizing yourself with the rest of the book. Understanding the basic elements and principles of TEM will greatly enhance your appreciation of the tremendous work, brilliance, and ingenuity of our predecessors.

Literature

[1]   De Broglie L. Recherches sur la théorie des Quanta. Ann. Phys. 1925; 10:22–128.

[2]   Davisson C, Germer LH. Diffraction of Electrons by a Crystal of Nickel. Phys. Rev. 1927; 30:705–40.

[3]   Thomson GP. Experiments on the Diffraction of Cathode Rays. Proceedings of the Royal Society of London. Series A, Containing Papers of a Mathematical and Physical Character. 1928; 117:600–9.

[4]   Knoll M, Ruska E. Das Elektronenmikroskop. Z. Physik 1932; 78:318–39.

[5]   Ruska E. Nobel lecture. The development of the electron microscope and of electron microscopy. Biosci. Rep. 1987; 7:607–29.

[6]   Haguenau F, Hawkes PW, Hutchison JL, Satiat-Jeunemaître B, Simon GT, Williams DB. Key events in the history of electron microscopy. Microsc. Microanal. 2003; 9:96–138.

[7]   Hawkes PW. Advances in imaging and electron physics: Growth of electron microscopy [Volume 96]. Academic Press; 1996.

[8]   Hawkes PW. The beginnings of electron microscopy. Orlando, San Diego, New York[etc.]: Academic Press; 1985. (Advances in electronics and electron physics, Supplement; vol 16).

[9]   Fujita H. History of electron microscopes. Int. Congr. Electron Microsc. 1986.

3 Construction of a transmission electron microscope and why you should know it

The resolution of the human eye is approximately 1 arc minute. This means that an average person, from a distance of 25 cm (the optimal viewing distance), can distinguish two objects spaced about 0.1 mm apart. If observed from a greater distance or if the separation between them is decreased, an illusion may arise, making it appear as a single object. Therefore, to perceive details separated by a smaller distance, additional tools must be used.

Magnification   ℹ

It will be crucial for us to grasp the concept of magnification. From an observational point of view (while using binoculars, a magnifying glass, or a microscope), magnification is the ratio of the size of the observed image to the actual size of the object. For example, if you observe an ant through a magnifying glass, and it appears to be 10 mm in length while its actual length is 2 mm, the magnification is 5×. Similarly, when photographing a microscopic image (both in light and electron microscopy) –magnification is the ratio of the size of the detail on the film to the actual size of an object.

During the production of prints, additional post-magnification was sometimes introduced, which was reflected in the image caption. The matter became more complex with the widespread use of digital detectors. We can still talk about magnification "in the plane of the sensor," but usually we observe this image further magnified on a computer screen or in print. The same photograph viewed on screens with diagonals of 14 and 24 inches will have different actual magnifications. Therefore, in the digital era, one should approach magnification measurements with caution and pay greater attention to the scale bar presented in the frame, as "the same magnifications" on devices from distinct manufacturers may represent entirely different "actual magnifications."

The simplest among these tools is the single lens (magnifying glass or loupe). It slightly magnifies an object, thus enabling the observation of finer details at the same angular resolution. An extension of this concept is the basic transmitted light microscope, composed of three lenses: condenser, objective, and eyepiece. The first lens, the condenser, focuses light rays from the source and directs convergent beam onto the observed object. Light interacts with the partially transparent specimen and reaches the objective, which magnifies the sample image 5× to 100×, and creates an inverted intermediate image. This image is further inverted and magnified additional 5–20× through the third lens, the eyepiece. By replacing single lenses with modern multi-lens systems, a cumulative magnification in the range of 1,000–2,000× can be achieved. Due to diffraction limit, the size of objects resolved using a light microscope can be approximated by the equation:

$$d = \lambda / 2A, \tag{1}$$

where d is the linear resolution (the ability to distinguish two points), λ is the illumination light wavelength, and A is the numerical aperture of the objective. For objec-

https://doi.org/10.1515/9783111317014-003

tives working in air, A can have a maximum value of 1.0 (usually between 0.1 and 0.9), and for immersion objectives (which require placing a drop of immersion oil with a specific refractive index between the specimen and the lens), this value can reach 1.4–1.5. Substituting $\lambda = 400$ nm (blue light) and $A = 1.5$ into eq. (1), we obtain a maximum resolution of approximately 133 nm. In practice, the real value rarely reaches 200 nm. This limitation was the primary motivation for developing observation methods using mediums other than light.

A good candidate to replace a photon is an electron, a small elementary particle with a nonzero mass and a negative elementary charge. However, the use of electrons for microscopy purposes requires the ability to overcome a series of challenges.

The first challenge is the generation of the electron beam itself. While for light microscopy purposes, even sunlight can be used (as in school microscopes with a movable concave mirror serving simultaneously as a condenser) or simple light bulbs or light-emitting diodes (LEDs). In the case of electrons, we must entrust this task to more complex systems. The simplest example of electron emission system is the one utilizing thermal emission phenomenon. Its essential component is a cathode – a thin wire made of tungsten (which combines a high melting point with a relatively low work function equal to 4.5 eV), which, under conditions of reduced pressure (used to prevent oxidation), is heated by the flow of electric current to a temperature of about 2,800 K (Figure 3.1a). The emitted electrons are then attracted to the positively charged anode (to which an accelerating voltage of around 1 to 300 kV is applied) and collimated by an electrostatic lens called the Wenelt cylinder. It is a cover with a hole surrounding filament, as seen on Figure 3.1b, and having a slight negative charge. At the exit of electron gun, a beam is divergent and its energy depends on the applied accelerating voltage (in this case, ranging between 1 and 300 keV) [1].

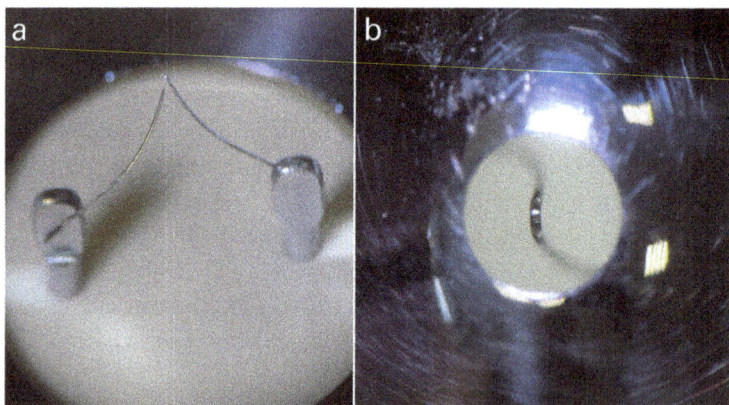

Figure 3.1: Parts of a thermal emission electron gun: (a) tungsten filament emitter, connected to supporting pins on insulating ceramics. Pin distance is about 10 mm. (b) Top view of central hole of Wenhelt cylinder with centered tungsten filament tip in the middle.

It is important to note that for a voltage of 80 kV, the speed of electrons will be 0.5c (half the speed of light), for 120 kV about 0.6c, and for 200 kV about 0.7c. This implies that in theoretical considerations and calculations, relativistic effects should not be ignored. Figure 3.2 compares the electron beam velocities concerning the speed of light c and the wavelength λ under the same conditions. It is easy to observe that even at relatively low voltages, the electron beam exhibits a wavelength below 10 pm, a value 40,000 times shorter than the blue light mentioned above [2]. So we proved that the electron is a suitable candidate for our new imaging medium. We also know how to generate it. However, a crucial point is that the described electron emission system cannot operate under standard conditions. The tungsten cathode would quickly burn out, and the generated electrons would collide with the air molecules. Therefore, another challenge is to provide an appropriate environment for the operation of an electron microscope, which is a high vacuum.

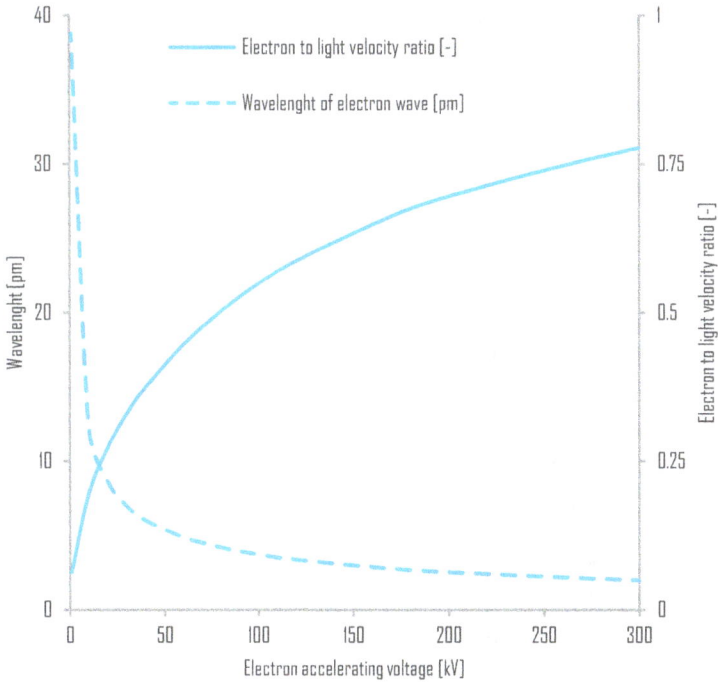

Figure 3.2: Electron-to-light velocity ratio and wavelength λ of the electron as accelerating voltage function.

According to eq. (1), light with a wavelength of 400 nm, along with a numerical aperture of 1, allows for achieving a resolution of about 200 nm. The electron beam for a voltage of 200 kV exhibits a wavelength of only 2.5 pm, but a conventional transmission electron microscope achieves a resolution in the range of 120–210 pm. Therefore, is it justified to further increase the voltage to shorten the wavelength of the electron beam?

Not really, because in TEM most of the resolution limits comes not from wavelength, but from aberrations in the systems (which require the use of low A in the eq. (1))

While light rays in typical light microscopy samples propagate through gaseous media (such as dry air), liquids (water, immersion oils), and solids (glass, thin tissues) with minimal scattering, electrons are much more easily scattered or absorbed. The free path of an electron at atmospheric pressure is measured in tens of nanometers, making observation impractical. Therefore, we look for a situation where an absolute vacuum is inside our microscope. Unfortunately, since vacuum is not natural in Earth's environment, we are forced to accept an available compromise. The simplest thermal emission systems (as described above) can operate at pressures (vacuum) around 10^{-2} Pa. This is an order of magnitude lower pressure than the pressure within the Kármán line, located 100 km above the Earth's surface and symbolically separating the atmosphere from outer space. An unsurpassed example is ultra-high vacuum microscopes (UHV), with vacuums reaching 10^{-8} Pa, values similar to those achievable near the moon. While UHV microscopes often offer unique opportunities for direct observation of material synthesis during observation and avoid its harmful interactions with gases (such as surface oxidation), typical electron microscopes operate at higher pressures, where demand (benefits of achieving a specific vacuum) is balanced by supply (cost of a given solution). For this reason, the simplest thermal emission microscopes are often equipped with simplified vacuum systems and sophisticated pumping systems are the domain of more demanding types of electron emission (discussed in the next chapter). Regardless of the required vacuum level, at least three types of pumps can be distinguished:
- Preliminary pumps (oil rotary – Figure 3.3a, membrane, scroll),
- High-vacuum pumps (turbomolecular – Figure 3.3b, diffusion – Figure 3.3c),
- Ultra-high vacuum pumps (ion – Figure 3.3d, sublimation, cryogenic).

Essentially, only the first group can operate at atmospheric pressure, while the others require the establishment of a preliminary vacuum. For this reason, preliminary pumps are an integral companion to more advanced vacuum pumping methods. They are particularly noticeable because their vibrations often require placing them outside the microscope. The simplest electron microscopes utilize a system comprising a preliminary oil-rotary pump and a high-vacuum diffusion pump. The advantages and drawbacks of each vacuum system will be discussed in greater detail in the next chapter as they can be crucial in planning and performing experiments [3, 4].

Figure 3.3: Different types of vacuum pumps: (a) two-stage oil rotary pump, used for rough pumping, (b) oil-free, high-vacuum turbomolecular pump with connected water cooling, (c) high-vacuum diffusion pump with visible water cooling pipes, (d) ion getter pump, mounted on TEM.

We have reached a situation where the generated electrons are moving in the high-vacuum environment of the microscope. Now, the next essential elements of our microscope will be the lenses. Earlier, we discussed the basic configuration of a light microscope, consisting of a condenser lens, an objective lens, and an eyepiece. The same configuration accompanied one of the earliest electron microscopes (the E. Ruska microscope from 1933, which surpassed the resolution capabilities of the light microscope, and was built with three lenses) and remains relevant to this day. Of course, the entire design has undergone justified development – a modern condenser system consists of 2 to 4 lenses, the objective has two separate pole pieces (and sometimes an additional pair of long-focal-length lenses, known as mini-lenses or Lorentz lenses for observing magnetic structures), and the projector comprises from 4 to 8 lenses (sometimes separated into intermediate lenses and projection lenses). These adjustments aim to minimize aberrations, achieve rotation-free imaging during magnification changes, and gain better control over the beam in various imaging modes. However, the idea of constructing electron lenses remains unchanged and can be divided into two basic groups: electrostatic (electric) lenses and magnetic (electromagnetic) lenses [5].

The former consists of three coaxial electrodes with apertures, where the outer plates are set to the same (zero) ground potential, and a high negative voltage is applied to the central plate. Such lenses were popular in the early years of electron microscopy. However, because of the necessity of applying high voltages (often reaching up to 50 kV), the risk of electrical breakdown, and the need to maintain electrode cleanliness, they did not survive the competition with magnetic lenses. A variation of the electrostatic lens is the Wenthel cylinder. Interestingly, electrostatic lenses found their place in the construction of a low-voltage transmission electron microscope (LVEM) projection system by Delong Instruments. However, there is no doubt that the vast majority of electron focusing on a global scale occurs with magnetic lenses. They consist of an external coil assembly located outside the vacuum through which a relatively high current flows, measured in single amperes. Therefore, they require water cooling. The coil windings are surrounded by the lens magnetic system, which on the vacuum side connects to the magnetic pole piece, usually in the shape of a cone with an internal through hole. This allows for continuous control of the lens focal length by changing the current flowing through the coil. Unlike optical lenses, changing the focal length of the system does not require replacing the entire lens or eyepiece, only adjusting the currents flowing through the lenses. In practice, this is performed by a control system (formerly vacuum tube, later transistor, and now digital), giving the user the ability to choose the magnification (by controlling the projection system), move the focal plane relative to the sample (using the objective system), and illuminate the selected area of the specimen with the required intensity of the electron beam (thanks to the condenser system). This brings our hypothetical microscope design closer to functionality.

We are left to address the final limitation. Although in the case of visible light, we were equipped with a relatively effective and maintenance-free detector (our eyes!), electron detection is a more demanding. Historically, the first method for recording electronograms (electron images) was exposure to a photographic film, which darkened under the influence of electrons similarly to light exposure. Over the decades, photographic films have, in some respects, surpassed digital detectors. Even in the early twentieth century, some electron microscopy laboratories were equipped with darkrooms, a practice that continues today. Nevertheless, while the photographic film is suitable for capturing images, it is necessary to have a system that allows observing the experiment's result in real time. The appropriate frame, precise focus, and refinement of the imaging parameters are essential. For this purpose, luminescent screens (also sometimes called fluorescent or phosphor screens) were used in the early years of electron microscopy. They consist of a thin layer of scintillating or cathodoluminescent material, such as ZnS or CdS. The luminescent screen was (and in some models still is) directly observed by the operator through a metalized inner window, as shown on Figure 3.4a. The screen can be tilted to an angle to facilitate focus using a low-magnification microscope. In many modern designs, the screen is not accessible to the operator but is observed using a light camera, with the signal transmitted to the microscope control environment. The idea of converting the electron signal into light using a scintillating material also forms the basis for the construction of conventional CCD/CMOS electron detectors. This camera is positioned below the fluorescent screen and is also coated with a scintillating substance (Figure 3.4b). The resulting light is transmitted below using a honeycomb-structured set of optical fibers (visible in the case of too intense image exposure) and reaches a conventional CCD/CMOS matrix sensitive to the generated light. Sometimes the image is taken outside the vacuum using a mirror and an optical system, and collected with an external light camera (Figure 3.4c) In the most demanding applications, direct electron detection cameras are used, but they usually cooperate with a conventional and less delicate CCD/CMOS camera discussed above.

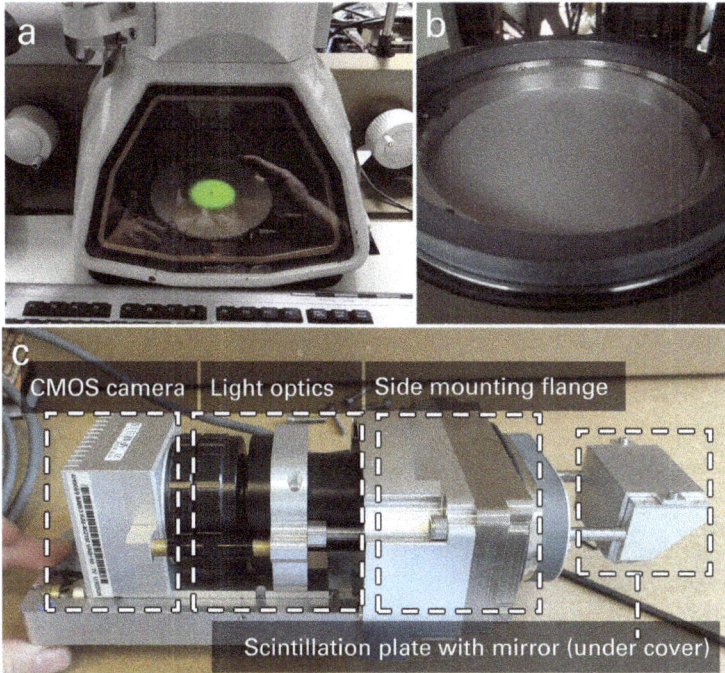

Figure 3.4: Some electron-detecting devices: (a) fluorescent screen with typical, green emission, seen in projection chamber of JEOL JEM-1200EX microscope, (b) scintillation plate of EMSIS Quemesa electron camera, CCD sensor placed below plate, not visible, (c) side-entry EMSIS Veleta camera without cover, with marked main parts.

In this way, we have preliminarily discussed the structure of the transmission electron microscope in an analogy to a simple light microscope. We successfully generated an electron beam, transmitted it in a vacuum environment to the optical system, and after interacting with the specimen, directed it toward available detectors. This simplified and idealized setup will form the basis for further consideration. In the following chapters, we will discuss in sequence the impact of the microscope's design and the specimen on the issues related to effective electron observation.

---

i Take-home messages

– One can easily compare the construction of a transmission light microscope and a transmission electron microscope by examining the radiation sources (light or electrons), the observation environment (diverse in light microscopy and vacuum in electron microscopy), the lens structure (glass versus magnetic and electric lenses), and the detection method (electron microscopy requiring the transformation of electron signals into light).

– The simplest source of electrons is a tungsten filament heated to glow. However, electrons do not exit a conventional light bulb because they are not adequately accelerated by an external potential, and the glass envelope effectively absorbs them.

– The vacuum in an electron microscope ranges from 10^{-2} to 10^{-8} Pa, equivalent to the conditions in space between Earth and the Moon.

– Electrons in an electron microscope are manipulated using magnetic or electrostatic lenses.

– The characteristics of electron sources, lenses, and the vacuum system can be crucial in the context of observations. We will expand on this topic in the next chapter.

Literature

[1]   Williams DB, Carter CB. Transmission electron microscopy: A textbook for materials science. 2nd ed. New York: Springer; 2008.

[2]   Reimer L, Kohl H. Transmission electron microscopy: Physics of image formation. 5th ed. New York NY: Springer; 2008. (Springer series in optical sciences; vol 36).

[3]   Yoshimura N. Vacuum technology: Practice for scientific instruments. Berlin: Springer; 2008.

[4]   Chambers A. Modern vacuum physics. Boca Raton: Chapman & Hall/CRC; 2005. (Masters series in physics and astronomy).

[5]   Hawkes PW. Topics in Current Physics: 18 Magnetic Electron Lenses. Springer-Verlag; 1982.

4 Know your microscope – the impact of the microscope design on research capabilities

The construction of a transmission electron microscope can have a significant impact on your work. Here, I discuss the most important factors in the order of discussion from the previous chapter: we will start with different types of electron emission and then discuss the features of vacuum systems, the construction of electron lenses, and go onto different types of signal detection [1].

In terms of electron emission, we can categorize transmission electron microscopes in two ways: in terms of achievable accelerating voltages and by the type of emission. From the point of view of accelerating voltages, we can mention the following types of devices:

- Low-voltage electron microscopes (LVEMs) – the accelerating voltages of this type of devices are much lower than 80 kV, and in the case of commercially available devices, they are in the range of 5–25 kV. Delong Instruments stands out in the production of this type of construction (Figure 4.1a). Company has devoted itself to simplifying the construction of a transmission microscope so that it fits on a desk and does not exceed the dimensions typical of scanning electron microscopes (SEM). For this purpose, a number of interesting solutions have been introduced such as lenses based on permanent magnets or additional magnification achieved using a light microscope system. They are difficult to recommend for advanced materials applications, but due to the significant scattering and low energy, electrons can offer high contrast in basic biological and nanomaterials applications.
- Microscopes dedicated to biological applications – they are characterized by maximum voltages in the range of 80–120 kV and are often the first choice for units specializing in biological imaging of ultrathin tissue sections or negatively stained samples. They can also be useful in diffraction studies and everyday work in the field of material imaging. However, one should be careful because some of these types of devices have deliberately limited functions that are less useful during biological imaging. For example, they may not be equipped with a diffraction aperture and the objective aperture may have a limited number of diameters. These types of devices include a number of historical TEMs, such as the Zeiss EM900 family of devices (with a characteristic orange column, as shown in Figure 4.1b), and also modern designs offered by JEOL, Hitachi, and ThermoFisher Scientific (TFS, formerly FEI and Philips). For simplicity, these types of microscopes are usually equipped with thermal emission (W or LaB6) and a not very demanding vacuum system, which, however, does not reduce their research value.
- Microscopes dedicated to high-resolution imaging – they are characterized by a maximum voltage in the range of 200–300 kV. They are often, but not always,

https://doi.org/10.1515/9783111317014-004

equipped with field emission or field-thermal Schottky guns as well as spherical aberrations correction systems. Some of them (TFS Krios, JEOL Cryo-ARM) are dedicated to high-throughput cryogenic analyzes (Cryo-EM) or automatic semi-conductor metrology. From the point of view of this study, the most interesting are the universal devices or devices dedicated to material imaging applications.

Figure 4.1: Voltage-size dependence for TEMs: (a) 5 kV table-top microscope Delong Instruments LVEM 5, (b) 80 kV, biological Zeiss EM900, and (c) 200 kV, universal JEOL JEM-F200.

As a rule, the greater the accelerating voltage, the greater the TEM itself (Figure 4.1). However, the height of the column can also be significantly increased by systems correcting spherical and chromatic aberrations. In turn, in recent years, there has been a trend of enclosing devices in housings without access to the fluorescent screen (Figure 4.2). Because the screen no longer has to be at the height of the operator's desk, the entire structure can be slightly lower (which could be important when looking for a place for a new installation).

---

i Do not be a slave of your accelerating voltage

It should be noted that the accelerating voltage of the device does not actually limit the ability to generate data, but is simply one of the features of the device adapted to a specific configuration. In the literature, one can find excellent material research performed at voltages of 100–120 kV and imaging of thin tissue sections at voltages of 200–300 kV. However, it is worth being aware that for specific applications, manufacturers deliberately select the most useful ranges of accelerating voltages. And let's not forget that in many cases we can also work with lower than maximum and optimum accelerating voltage.

Another important piece of information linked to the electron gun system is the type of emitter and the nature of the emission. With some simplification, we can divide them into systems using thermal emission (based on tungsten – W or lanthanum/cerium hexaboride LaB6/CeB6) and field emission (Schottky emitter, which is basically a field-assisted thermal emitter and an emitter with cold field emission) [1]. For each of the above-mentioned types of emitters, the following increases: purchase and operating

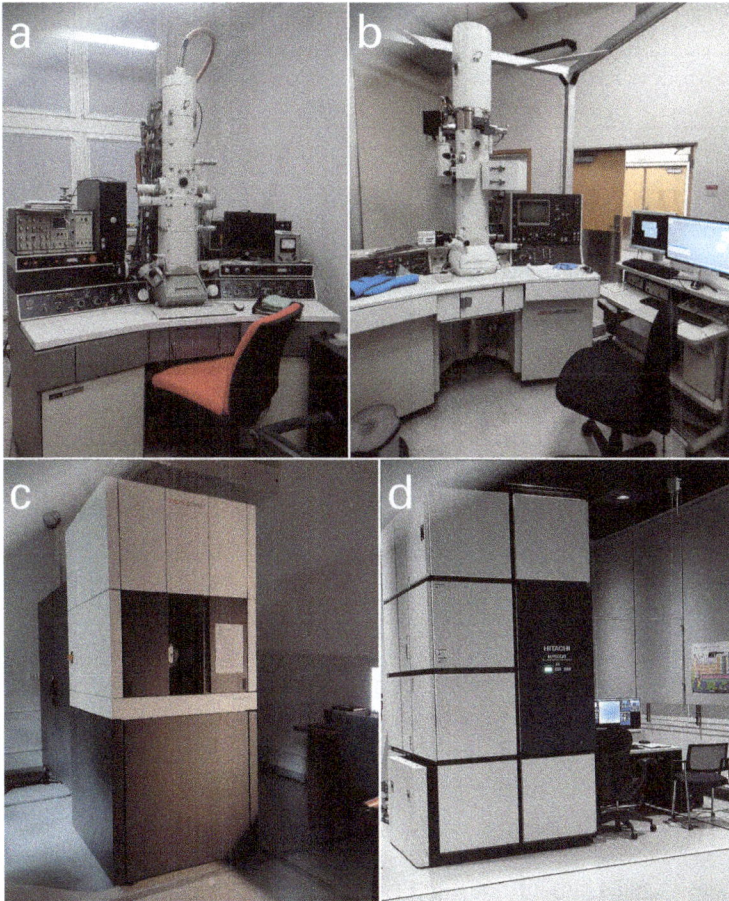

Figure 4.2: Fifty years of development in the design of TEM microscopes dedicated to material research: (a) universal JEOL JEM-100C with thermal emission, 100 kV, manufactured in the 1970s, (b) JEOL JEM-2010F with Schottky emission, 1990s, (c) ThermoFisher Scientific Talos F200i, 200 kV with Schottky emission, 2020s, (d) Hitachi HF5000, 200 kV with cold-field emission and probe corrector, 2020s.

costs, vacuum requirements, and the obtained emission brightness, and at the same time, the energy dispersion of the beam and apparent diameter of the source decrease (which are very positive features). It is therefore possible to roughly compare the described electron sources as shown in Table 4.1.

The table allows you to easily read the main differences between emitters. The tungsten-based thermal gun is the simplest and has the lowest vacuum requirements, which lead to the overall simplification of the microscope design and its relatively low price. This type of gun often allows obtaining subnanometer resolutions, but a very wide crossover size and high unfavorable energy spread lead to difficulties, for example, in high-resolution scanning transmission imaging (STEM). This type of emission is very

Table 4.1: Relative comparison of the most important features of different electron guns.

Emitter type	Thermomix	Thermonic LaB$_6$/CeB$_6$	Schottky W/ZrO	Cold field
Emission type	Thermionic electron gun		Field emission gun	
Vacuum needed [Pa]	10^{-2} to 10^{-3}	10^{-4}	10^{-6}	10^{-8}
Practical lifetime	100 h	1,000 h	3 to 5 years	1,000 flashes
Brightness	+	++	++++	+++++
Emission stability	+++	+++	+++++	+
Energy spread (eV)	+	++	++++	+++++
Beam coherence and the possibility of high-resolution imaging	+	++	++++	+++++
Ease of use	+++++	++++	+++	++
Price attractiveness	+++++	+++++	+++	++

popular in entry-level SEM and basic or slightly older TEMs. Low vacuum requirements mean that a diffusion or turbomolecular pump is enough to obtain sufficient vacuum, without advanced and expensive ion getter pumps. Microscopes with a tungsten emitter can be permanently turned off and turned on only for observations. Theoretically, the cathode requires replacement approximately every 100 h of operation, but if appropriate and delicate operation is maintained, results of up to 1,000 h can be achieved. The replacement of the cathode itself, including the necessary cleaning of the working area and re-pumping high vacuum, can take less than an hour. It is therefore not surprising that microscopes with a tungsten emitter that are few years or several decades old still remain a durable and cheap-to-use "workhorse" in many laboratories. They are also a great choice if you plan to build and test your own experimental accessories. Even in the event of a sudden vacuum failure while the emission is on, it is difficult to do more damage than a burnt emitter, which is relatively easy to replace cheaply and quickly.

A development of the tungsten emitter is a gun using a cathode based on LaB$_6$ or CaB$_6$ crystal. Very often, these types of emitters are interchangeable with tungsten emitters, requiring only a slight change in the electrical parameters of the system (using a switch). It also happens that this type of emitter can be purchased as an accessory replacement for the tungsten thermal gun. It has many benefits – greater brightness, smaller source diameter, and smaller energy spread. A LaB$_6$ or CaB$_6$ emitter approximately provides approximately 10× greater lifespan and at approximately 10× greater cost. However, there is one important limitation. The LaB$_6$ or CaB$_6$ emitter also requires at least an order of magnitude better vacuum than the tungsten emitter. For this reason, microscopes of this type have slightly more complex vacuum systems (also equipped with ion pumps, absent in microscopes with tungsten emission), and if the vacuum system is still simplified, it requires continuous operation, without turning off the microscope every day. It is also not as resistant as the W emitter to vacuum deterioration or heat shocks. Therefore, it requires a bit more attention during everyday maintenance.

Field emission guns are a significant step up in quality compared to thermal emission guns. The most popular of them are Schottky emitters, which are basically thermal emitters aided by a field effect. This type of emitter is a sharp, single-crystalline tungsten tip, coated with ZrO_2 to reduce the electron work function. The loss of ZrO_2 is refilled from a reservoir located further up the emitter. The Schottky gun is characterized by much greater brightness, a more favorable and small emission diameter and energy spread, as well as a stable emission value, which, if necessary, can reach relatively high values. Emitters of this type are quite universal. The most demanding systems are sometimes equipped with an additional monochromator, further reducing the energy spread for analytical purposes. However, a system with a Schottky emitter requires a vacuum at least two orders of magnitude better than that required for LaB_6 or CaB_6 emitters. This complicates the vacuum system, increases the compexity of the entire system and increases its price. In return, the electron gun offers several years (usually 3, but some systems could operate continuously for over 10 years) of stable operation. Under regular work, the emission is usually activated 24/7, which means that even a device that is little used will require regular replacement of the emitter, which is more expensive than in the case of the previously discussed systems. In the event of long-term interruptions in work, it is possible to turn off emissions, but remember that a high vacuum should, in principle, be maintained continuously (it is relatively easy to maintain, but rebuilding it is very time-consuming and often problematic).

The last type of popular electron gun discussed is the so-called cold field emission (cold field gun). It has the highest vacuum requirements (which means that the gun comes from the factory with a ready-made vacuum, which is only maintained by a microscope, and the entire gun modules are replaced during service), the most favorable energy spread and beam coherence parameters. Ultra-high vacuum (UHV) is required to slow down the contamination of single crystal W as much as possible. To obtain the possibility of field emission at room temperature, the emitter is heated to high temperatures for a short while before each operation to clean its surface (this is the so-called flashing procedure). At this point, it reaches the maximum of its emissions, which slowly decreases over time until it reaches zero. For this reason, the emitter must be reheated every few hours of operation. Changing the emission value can be problematic during long-term experiments and automated tests (such as electron tomography), but there are methods available to deal with this difficulty. Cold field emitters are long-life and their lifespan is limited by the amount of flashes. With daily work, they should not survive less than 3 years, and practice shows that they can actually work for far more than 10 years. Unfortunately, the price of this type of system, its vacuum system, and replacement costs is the highest in all the mentioned types of guns. However, in modern high-class microscopes we usually find a cold field emitter or a Schottky emitter (often with an additional monochromator).

As you can see, as the imaging capabilities of advanced types of emissions increase, the price, complexity, and vacuum requirements of the system grow as well. In the previous chapter, we roughly discussed the division of vacuum pumps from the point of view of the vacuum obtained (pre-pumps for low vacuum, and high-, and ultra-high-vacuum pumps). Each of the types of emissions described above requires the use of the first two types of pumps. The pre-pump provides operating conditions for a diffusion or turbomolecular pump, ensuring sufficient vacuum for the operation of the tungsten emitter, and sometimes even the LaB_6 or CaB_6 emitter. Both of the above-mentioned types of vacuum pumps can be divided into two types – oil and oil-free systems [2, 3]. Oil pumps (rotary oil pump for rough vacuum and high-vacuum diffusion pump) are almost maintenance-free, and ensure many years of stable operation. Instead, they may end up polluting the microscope environment with vacuum oil fumes. Special oil traps are used to limit this effect, but there is no doubt that there are research areas that do not prefer pumps of this type. They may disturb experiments involving the deposition of materials in situ and accelerate the contamination of the sample during imaging. In turn, oil-free pumps (membrane or scroll pumps for low vacuum and high-vacuum turbomolecular pumps) offer a cleaner working environment, in return requiring quite regular servicing and greater maintenance care. However, the division into oil and oil-free pumps should not be treated as an oracle. The author also worked on systems where a modern microscope with cold field emission was equipped with two oil pumps, and an old microscope with tungsten emission operated for over 20 years in an unmaintained turbomolecular pump. Also, UHV systems can use oil rotary pumps, but they are turned on a few moments later than the turbomolecular pumps, which prevents oil vapors from entering the vacuum chamber. However, reaching the ultrahigh vacuum, it is worth mentioning that the most popular pumps are used to obtain it. These are ion pumps (ion getter pump, ion sputter pump, IGP), which are oil-free pumps with no moving parts. In an ion pump, a high vacuum (which still contains unwanted gas molecules) is ionized using electric and magnetic fields. The resulting ions hit the titanium cathode, sputtering titanium atoms. Then the atomized titanium atoms create a fresh layer susceptible to the adsorption of gas molecules. As a result, active gases (hydrogen, oxygen, carbon monoxide, etc.) are adsorbed on the getter layer, and inert gases (argon, etc.) are ionized and adsorbed on the cathode [4]. Pumps of this type make it relatively easy to achieve the operating conditions of field emission guns, but they pump water vapor quite poorly. For this reason, many systems periodically turn off the ion pumps and return to the high-vacuum pumps to clean the column of water vapor and some of the remaining impurities.

When talking about vacuum systems, it is impossible not to mention a very popular accessory that essentially functions as a cryogenic vacuum pump. It is an anticontamination device (ACD), built as a cage surrounding the sample inside the objective pole piece (Figure 4.3a). Cooled to liquid nitrogen temperatures, it adsorbs residual gas molecules or molecules that have been removed from our sample during imaging, for example, because of electron beam interaction. The main task of ACD is to slow down the process of sample contamination, the build-up of a layer of amorphous substances in the observation or surrounding area (Figure 4.3b), which makes observations difficult and reduces image contrast. In some imaging modes (e.g., STEM and long-term EDS map imaging), the use of ACD is essential. Importantly, the capabilities of this accessory are limited, and working with a cooled ACD for too long will cause it to no longer fulfill its function. For this reason, at the end of the working day (and in some systems even on the weekend), we stop cooling the ACD and heat it to the ambient temperature. This is accompanied by a significant deterioration of the vacuum, so we turn off the ion pumps in advance so that a large amount of impurities can be pumped out by more resistant high-vacuum pumps. However, one do not have to worry about a complicated procedure – it is usually automated, and you will get all the important information from the user manual or your more experienced colleagues.

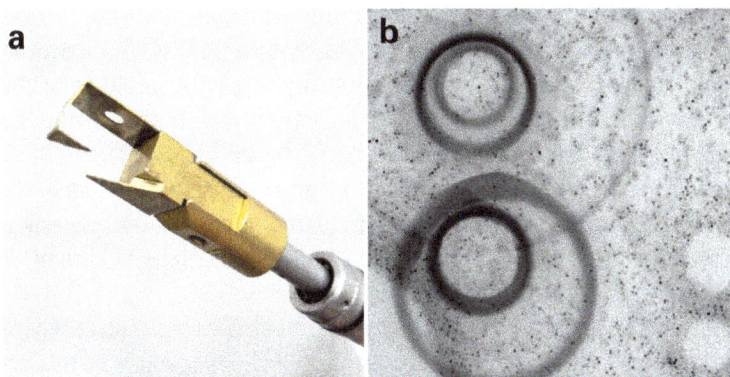

Figure 4.3: (a) Gold-coated anticontamination device removed from the TEM column with the bottom cut for the objective aperture, (b) example of overlapping contamination rings surrounding the areas illuminated previously by electron beam.

If we have already discussed the important features of the electron gun and the associated vacuum system, on our way down the column we will encounter optical elements – lenses (focusing electrons), deflectors (moving and centering the beam) and stigmators (responsible for correcting astigmatism). While the details of their work are usually not of interest to the operator, it is worth remembering that any sudden thermal changes occurring in the system reduce its stability and make it difficult to obtain the best results. For this reason, we can expect that immediately after inserting a new sample, changing the temperature in the laboratory (due to the visit of a large group of people or through the air flow), or a significant change in the current flowing in the lenses (which occurs when switching between the lowest and medium magnifications), we will observe greater than usual image drift. In such a situation, the system can be left to stabilize or the time can be spent assessing different areas of the specimen and selecting the best locations for final imaging.

While the TEM samples themselves will be the subject of the next chapter, here we will discuss some important issues related to the sample holder. In the vast majority of popular devices, we deal with the so-called "side-entry" systems, where the sample is installed inside the lens from the side of the microscope, often just above the operator's head, through a goniometer (a device that positions the sample in space) and an additional vacuum lock. This type of configuration is not perfect to easily obtain the highest resolutions, but it offers a number of advantages. Virtually every holder can be tilted relative to the main axis, which facilitates diffraction studies, and the rod itself is a bridge between the sample and the environment around the microscope. This creates excellent opportunities for research using in situ microscopy. Therefore, you can buy specimen holders that offer, for example, resistive heating of the specimen (with heater and thermocouple wires led outside), its cooling (sample connected with a copper cable to an external liquid nitrogen tank), biasing (electrical connections), stretching the sample inside the column (mechanical system powered by an external motor), observation of liquid samples (by supplying liquid from the outside between two impermeable but electron-transparent membranes inside the column), and many others. Even if the holders do not support in situ research, they can have many other applications. For example, they may contain an additional tilt axis (double tilt holder), which facilitates diffraction studies and precise sample orientation for high-resolution TEM/STEM. Some of them can also be made of carbon or beryllium, which reduces the background signal in EDS analyses.

Another important TEM system is the projection chamber. This is the place where the fluorescent screen and/or electron cameras are installed. Since much of this arrangement was covered in the previous chapter, we will only add some practical details at this point. I expect that most readers use scintillation cameras with CCD/ CMOS detectors. They are relatively sensitive and at the same time have long-life. Quite naturally, they have a vignetting defect, i.e., the edges of the image are darker than its center. To obtain a clear image, calibration images are taken during the camera calibration stage. They are usually not important to the user, but it is worth being aware of them. The first one is an image without the electron beam falling on the detector. This way you can remove, for example, hot pixels. The second correction involves taking a photomicrograph without the sample in the field of view and with the beam evenly illuminating the camera. This allows you to hide vignetting, the image of the optical fibers connecting the scintillator and the matrix, as well as mask small dust particles and scintillator defects. If you see more still image noise or dust after longer operation, you may want to refresh the calibration process.

Protect your camera ⓘ

Importantly, cameras are also sensitive to excessive exposure to electrons. In an overexposure situation, the scintillator may require several moments to return to normal operation. If we observe a very contrasting image with high beam intensity (such as TEM gratings at very low magnification), it may turn out that the "ghost" of this image will accompany us for a few moments. In such a situation, it is not worth recalibrating but simply wait for the relaxation of the excited fluorophore. We also avoid excessive, spot exposure. For this reason, we do not center the focused beam on our high-resolution camera, but on an auxiliary fluorescent screen (in new systems also equipped with a digital camera). In the case of diffraction exposure, it is worth using an additional central (undiffracted) spot blanker, often called a beam stopper. It allows you to protect the matrix against long-term exposure of the central part of the camera to the strongest part of the beam. We will discuss this issue more precisely in Section 6.12.

The last piece of information we should remember is that the detector may be located in different places in the projection chamber. Usually, we can indicate two most popular positions – below the fluorescent screen (so-called "bottom mount") and above the screen (so-called "side mount" – because the camera must be equipped with a system for inserting and removing the image element. Awareness of the camera's position can be quite crucial. If we work with a "bottom mount" camera (which, in fact, are also old-fashioned exposure systems for photographic plates), it is obvious that in order to obtain an image or exposure on it, we must raise the fluorescent screen upward. It is often automatic or semiautomatic, but if a modern CCD/CMOS detector has been installed in older microscope, it may turn out that our manual intervention is required. On the other hand, if we are dealing with a "side mount" camera (Figure 3.4c), lifting the screen has no future – we need to physically insert the camera elements into the optical path. This is usually done using pneumatic or electric control. Even if we have turned off our detector, it may turn out that we see nothing on our fluorescent screen – in such a case, we must, of course, remove the camera elements blocking the electron beam. I deliberately do not write "detector" or "scintillator" because side-mount cameras could be built in two ways. Sometimes the detector is located directly under the scintillation screen and the fiber optic system (as in a classic "bottom mount" camera), and sometimes (even more popular) only the scintillation screen with a mirror placed under it is inserted. In this configuration, the glass optical system and the actual light detector remain outside the microscope and outside the vacuum. These types of configurations are generally less resolving and less sensitive, but they allow you to easily install a digital camera in most of old-type microscopes, without sacrificing the functionality of photographic films.

4.1 TEM optics working principles

The microscope optics itself consists of several or a dozen lenses, but they are always grouped into three basic groups: the condenser system, the objective, and the projector system. The schematic description of visible parts of TEM is shown in Figure 4.4.

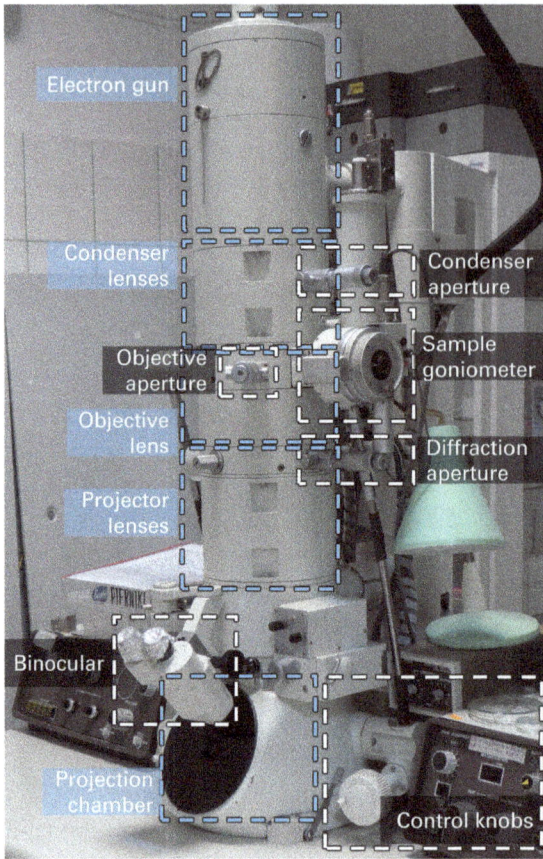

Figure 4.4: Position of some parts of a TEM, using the example of a vintage JEOL JEM-100S.

Using the condenser system (lenses, deflectors, and stigmatator), we are able to regulate the number of electrons falling over the specimen (using the spot size parameter related to the current flowing through the first condenser lens), the diameter of the spot (coinciding with the observation area), and the angle at which the beam cross the sample (often close to a 90° angle, but not always), and the convergence of the beam (it may be parallel, convergent, or divergent). The simplest control elements of the condenser system are the Intensity knob (adjusting the convergence and focus of the beam) and the pairs of Beam Shift and Beam Tilt knobs (one for X and the second for Y direction). Intensity and Beam Shift are used almost constantly during normal work because they help us ensure that the illumination area of our sample corresponds to our magnification. It is worth noting that our shift operates in a very small area and the main selection of a place on the sample is done by mechanically moving the sample relative to the microscope using a goniometer and the built-in stepper motors, gears, and sometimes a piezoelectric system.

The objective system (consisting of the most important component in the entire system: the objective lens, the objective stigmator, and side systems) is responsible for the interaction of electrons with the sample and the transfer this information down the microscope. We use objective lens to move the focal plane relative to the sample using the Focus knob. While in most microscopic methods we aim to find an in-focus position; in the case of TEM we often take micrographs in a slight underfocus position (described in Chapter 6). In such a case, darker, so intensively scattering elements of our image (e.g., nanoparticles on the background of an almost transparent carbon film) gain thin light halos, somewhat similar to the graphical software "unsharp mask" filter. In contrast, we usually do not use the overfocus position because elements of our image become brighter and surrounded by a dark fringe. With many basic imaging methods, it is enough to select the plane of focus using your subjective assessment of the quality of the image, but if you want to image atomic planes, it is worth learning the concept of "Scherzer defocus." To put it simply, this is the underfocus position, depending on our acceleration voltage and the spherical aberration of the lens, in which it is easiest to observe the phase contrast from the planes. Some systems make it easier for the user to enter this position if the user first sets the "in-focus" position and start the appropriate button.

The projector system (sometimes divided into an intermediate system and a projector system) is responsible for magnifying the image created by the objective lens and transferring it to the detector plane. Taking advantage of the fact that we are able to freely change the current flowing through the lenses, we can choose between different magnifications and observe the diffraction image from the place of interest. This is also how we control the projector system by selecting the imaging mode and magnification. It is usually the user's responsibility to ensure appropriate interaction between the condenser and projector systems. If we work at low magnifications, we need to expand the beam using a condenser (Intensity knob) and center the beam (using a pair of Beam Shift knobs). If we use increasingly higher magnifications with the same condenser parameters, we will notice that the image becomes darker or more noisy. In such a situation, we need to focus the beam on the area of interest with the Intensity knob and re-center it with Beam Shift knobs. These types of quick corrections become automatic and unconscious after gaining some experience, but at the beginning they take up a large part of the working time. It is also described in more detail in Section 6.9.

We have already mentioned that electron lenses are inextricably linked to various types of optical defects. One of these defects is astigmatism. For the condenser system, it manifests itself in the fact that our beam is not round, but elliptical, and changes orientation when passing through the focus. In such a situation, the condenser astigmatism corrector (often called simply Condenser Stigmator or Cond Stig) should be adjusted so that the beam becomes round again, as described later in Section 6.4. It happens that in different positions of the Intensity knob our astigmatism is slightly different and, for example, if we have a round spot in conditions of high beam concentration, after its expansion, it becomes slightly elliptical. In such a situation, it is worth setting the astigmatism at beam sizes typical for the highest magnifications we need.

Objective lens astigmatism manifests itself slightly differently (corrected with the objective stigmator or Obj Stig command or knobs). Not long ago we said that underfocus position is accompanied by a light fringe around dark elements of the image, and overfocus position is accompanied by a dark fringe. In the case of an image without astigmatism and a perfectly round image element (a carbon film with round holes is used for calibration), these stripes have an even thickness around the entire hole border, and as they approach the in-focus position, they gradually disappear (as shown later on Figure 6.8). In the case of an astigmatic image, the fringes will have different thicknesses in different orientations (it is easiest to compare fragments of a circle separated by 90°) or even close to the in-focus position we will find fragments of the rim typical of underfocus and overfocus (Fig. 6.10). In the predigital camera era, the typical procedure was to set the astigmatism on the perforated film and use these settings as the baseline. Unfortunately, many real samples do not have areas or pinholes where it is easy to set astigmatism. Then, at most, the astigmatism was corrected locally, guided by the subjective feeling of image sharpness. The situation improved significantly when sensitive digital detectors became popular. Thanks to this, we are able to observe in real time not only the electron image but also its fast Fourier transform (FFT). We are able to observe concentric circles or ellipses called Thon rings. They fade away and become larger as you get closer to the in-focus position (Fig. 6.9) Using astigmatism correction, we correct the image so that the rings on the FFT become perfectly round (Fig. 6.11).

Another element of the microscope that is important for the effects of our work are movable apertures. They allow you to change the convergence of the beam, limit the number of electrons in the system (condenser aperture), and cut out parts of scattered or diffractive beams from the projector system, which allows you to control the scattering contrast and observations in the bright and dark field (objective aperture) as well as select the location of the diffraction analysis (diffraction aperture). The objective aperture also helps reduce the charging effect of the electron-sensitive sample. Any change in the moving aperture during observation may result in the appearance of additional astigmatism, requiring further correction. This phenomenon is more visible and more contamination products are present in the aperture. For this reason, systems with oil pumps are more sensitive to this inconvenience, and apertures generally require regular cleaning or replacement (especially those holes that we use regularly). Apertures should generally be properly centered during operation using a mechanical or electro-mechanical positioning system. The symptom of an uncentered condenser aperture will be nonconcentric expansion of the beam as the Intensity knob is operated, and for the objective aperture, excessive astigmatism, lack of visibility of the image, or some shadowing could be visible. It is also worth remembering that the smallest objective apertures limit the field of view at low magnifications, and in some modes low magnifications even require removing the aperture out the beam path. The diffraction aperture is located in the plane of one of the virtual images in the projection system and allows the selection of an area for electron diffraction. As a rule, it should also be centered, and the place of interest is selected by moving the specimen and not the aperture. However, it is worth knowing that the position of the diffraction aperture (also called selective) is less crucial than the condenser and objective aperture. Selecting a specific hole size and moving (centering) the apertures can be done manually (appropriate knobs on the aperture mechanism on the column) or electrically (controlled from the operator panel or software). It may happen that in the same microscope there are two different types of aperture shift, electric for one aperture, and manual for another. The principles and methods of centering individual apertures will be discussed in Chapter 6.

Take-home messages

- There are two basic types of electron emission: thermal (cathode made of W or LaB_6/CeB_6) and field (field-thermal Schottky emission and cold field emission). Each of the subsequent types of guns has higher vacuum requirements, a higher price and more problematic operation, but in return they offer greater brightness, better beam coherence, less energy spread, and longer life.
- We distinguish between low-, high-, and ultra-high-vacuum pumps including oil and oil-free pumps. Each of them has its advantages and limitations and can constitute the strengths and weaknesses of our TEM.
- If your microscope is equipped with an anticontamination device, it is usually a good idea to use it, with a final heating at the end of the working day or as directed in the user manual.
- The optical system of a TEM microscope is divided into a condenser, objective, and projection system. They are largely independent of each other and it is worth knowing their function and operation. This will be discussed in detail in Chapter 6.

Literature

[1]   Williams DB, Carter CB. Transmission electron microscopy: A textbook for materials science. 2nd ed. New York: Springer; 2008.
[2]   Chambers A. Modern vacuum physics. Boca Raton: Chapman & Hall/CRC; 2005 (Masters series in physics and astronomy).
[3]   Yoshimura N. Vacuum technology: Practice for scientific instruments. Berlin: Springer; 2008.
[4]   Yoshimura N. Historical Evolution Toward Achieving Ultrahigh Vacuum in JEOL Electron Microscopes. Tokyo: Springer Japan; 2014.

5 Know your sample – selection of the preparation method according to your needs

When considering the preparation of TEM samples and the different approaches that can be used, we should start with the question: what do we actually require from our samples? I've listed some of the most important requirements below, but don't treat them as a list from most important to least important; rather, consider them as a set of connected and equivalent concepts.

First, the sample must be partially transparent to the electron beam. It is the diffraction and scattering of electrons in contact with matter that allows us to make observations. Various observation techniques may prefer areas with a smaller (10–50 nm) or larger (50–200 nm) thickness, but usually a rational limit is around 100 nm thick. This is dictated not only by the fact that thicker areas significantly limit the transmission of electrons but also by the fact that in TEM in the vast majority of cases we observe a two-dimensional (2D) image (projected onto a fluorescent screen or detector). The thicker the sample, the more our 2D image represents the larger volume (3D) of the sample, so the structure components we are interested in may overlap and be more difficult to analyze. This issue is clearly visible when we observe an aggregate of many small and transmissive nanoparticles. When they form a complex and multilayered structure, it is difficult to determine the exact shape and size of the individual components. The same limitation may manifest itself if we observe an area of thin foil that is too thick. An example of difficult observations of this type may be a situation in which individual crystallites of a polycrystalline substance will be clearly finer than the local thickness of the sample. Then, for obvious reasons, we lose the opportunity to study individual grains of our sample.

Second, the thickness of our sample should not vary much over a transparent area. If the sample surface is uneven or wavy (e.g., due to excessive stresses introduced during preparation), a large part of the image may be occupied by bending contours, obscuring valuable structural information. Of course, we can tilt the sample to move the contours, but an even and smooth surface of the sample definitely makes the observations much easier [1].

Third, our sample should be representative of the material. While this may seem obvious, it should be borne in mind that, in principle, the TEM sample may (but does not have to) provide the researcher with areas counted in tens and hundreds of square micrometers. This is a much narrower area than in many other methods (even scanning electron microscopy). For this reason, when analyzing the microstructure of a metal alloy, for example, it is worth starting with observations using light microscopy, then scanning electron microscopy, and only finally using transmission electron microscopy. In this way, you acquire knowledge about the material on an increasingly finer

https://doi.org/10.1515/9783111317014-005

scale, without omitting any of them. Thanks to this, you will avoid the mistake that the author once witnessed. A case of cracking metallic castings was analyzed by researchers delighted with the possibilities of TEM, who wanted to analyze the density of dislocations in the material to describe the mechanism of material damage. The research was already underway when a more detailed analysis of the material under a light microscope revealed the presence of macroscopic casting defects in the form of a significant number of gas bubbles. However, such significant defects were not possible to detect using TEM methods, which focused on solid fragments of the material. With this example in mind, I recommend that we examine the matter "from the general to the detail."

Fourth, our sample should not be modified or damaged during the preparation process. Many of the preparation methods (pregrinding, electropolishing, ion polishing, focused ion beam (FIB)) can introduce undesirable artifacts into the material, which will be partially discussed later in this chapter. At the beginning of your journey with TEM observations, if something surprises you in the analyzed image, consider whether it may be due to the influence of the preparation methods on your sample.

Fifth, the sample should be stable in the TEM environment. For this reason, special procedures are required for microscopy of liquid samples (liquid cell TEM), as the vast majority of solvents evaporate and boil under high vacuum conditions. The electron beam also has an impact on the sample – many materials change their structure during even minimal contact with high-energy electrons [2]. Crystalline organic substances can amorphize so quickly that it is almost impossible to perform diffraction studies on them. Many materials can also change their structure during observation. For this reason, if you are trying to reproduce literature imaging, look for information about the maximum electron dose and dose rate in the work you are analyzing [3–5].

The next requirements are additional wishes and potential limitations. If you are analyzing a nonconductive powder, in certain situations, it may accumulate an electric charge, have a local electrostatic effect, and for example, separate from the substrate. Unless individual nano- or microparticles in the TEM column do not constitute significant contamination, it is generally worth minimizing such situations. In some laboratories, you may also encounter a situation in which the analysis of ferromagnetic samples is prohibited. This is usually due to the fact that the magnetic material inserted into the strong field of the objective lens is for some short moment subject of significant magnetic field gradient. This may result in the sample being torn out of the specimen holder and ending up inside the column. This is not a critical situation because a sample of this type usually stands upright on the edge of the pole piece of the objective lens, and thus worsening the overall astigmatism and the performance of the microscope, it still allows observations and patiently waits for the column to be vented and the sample to be removed. However, it is not without reason that we avoid this type of situations, downtime, and costs. If you suspect that your microscope and holder do not like ferromagnetic samples, first read the user manual and/or contact the manufacturer's representative. Sometimes the specimen holder is very delicate and holds the sample quite poorly, and limiting the possibility of working on magnetic specimens is the factory's recommendation. In many designs, the above limitation can be overcome by having the specimen moved in and out of the column in specific imaging mode, which turns off or minimizes the current flowing in the objective lens. This can be, for example, low magnification mode (LM) or Lorentz microscopy mode. If, despite everything, the manufacturer, the head of your laboratory, or the microscope operator is not willing to work with a ferromagnetic sample, you should look for another device or a different sample or technique.

To summarize this short introduction, I would like to point out that as your experience increases, you do not necessarily have to stick to all of the above requirements. A slightly bent specimen may be suitable for excellent imaging, a sample that is too thick may be sufficient enough to create simple illustrations, and the negative interaction of the electron beam with the sample could reveal an interesting structural transformation. I hope that at this stage of reading you have already noticed that electron microscopy requires knowledge of its requirements and limitations, but at the same time it provides opportunities to overcome some of them or even turn limitations into

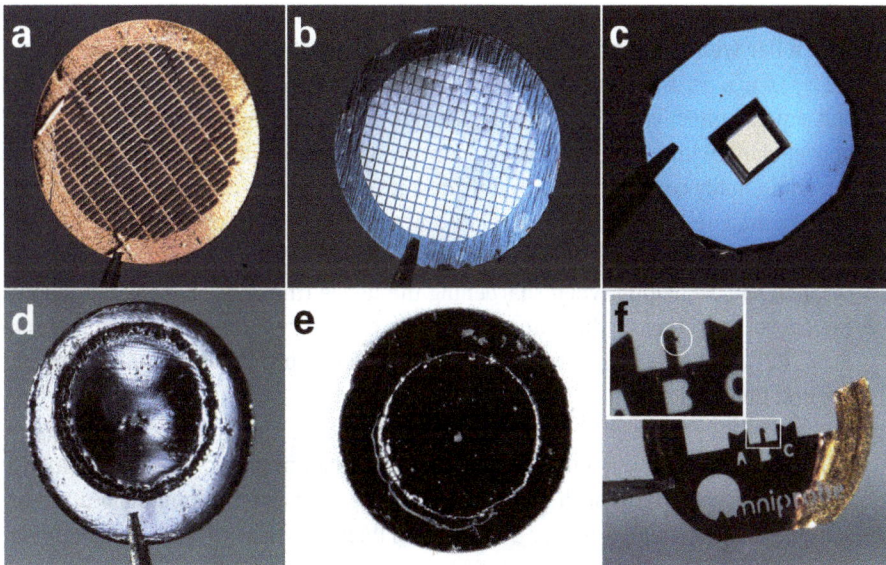

Figure 5.1: Different grids and samples for TEM, with a diameter of 3 mm for each sample: (a) metallic grid without additional film, (b) metallic grid with amorphous carbon film, (c) SiN film (yellow) on Si substrate, (d) metallic thin foil in reflected light, (e) metallic thin foil in transmitted light (perforation in the centre), and (f) FIB lamella mounted on copper FIB substrate (20 × 20 μm, lamella marked with circle).

advantages. In Figure 5.1 several sample substrates and TEM samples are compiled and are discussed later in the chapter.

5.1 Preparing TEM samples from nanomaterials

The simplest situation assumes that the material does not require special preparation. This applies, for example, to samples of nanomaterials in liquid solution. However, even in such an elementary situation, the researcher must be aware of the variables influencing the final effect. One of them may be the selection of an appropriate substrate. In the case of nanomaterials (and a number of other samples), we can use a whole range of different TEM substrates. These include

– Metallic grids without additional supporting film – The most popular ones are made of copper, but you can also buy commercial ones made of nickel, molybdenum, gold, or platinum. It is worth making sure that your material (especially the sample medium—solvent) does not react with the substrate material. Uncoated meshes can be beneficial in the case of material in the form of massive but thin flakes (though this is not the best substrate for studying 2D materials), fibers, samples after cutting on an ultramicrotome, or as a base for self-made coated meshes (Figure 5.1a). We usually use grids with a mesh size from 50 to 400. For example, the most popular mesh, 200, can offer window sizes of approximately 90 × 90 µm with a bar width of approximately 35 µm.

– Metallic grids with a thin film of amorphous carbon – These meshes (like some subsequent ones) share potential substrate material with the meshes described above, along with all the associated consequences. Unlike the previous ones, these are additionally covered with a thin (from 3 to 5 nm, though there are films of much greater thickness) and continuous layer of amorphous carbon (Figure 5.1b). This layer produces a relatively small background signal, typical for amorphous substances, and can be helpful when sharpening the image (the amorphous region allows the characterization of the defocus measure by observing the fast Fourier transform, FFT, of the image). If you have no idea what substrate to use, this is usually the best initial choice.

– Metallic grids with a thin film of Formvar and optionally amorphous carbon – This type of substrate is more resistant to cracking and is often used in complex preparation procedures such as negative staining or long-term immunolocalization staining of biosamples. It is very easy to make them on your own, and the additional coating with a layer of carbon (applied either at the factory or independently in a vacuum sputtering machine) makes the film more resistant to the effects of the electron beam.

– Metallic grids with a thin film of holey supporting film – Both carbon and Formvar substrates can be equipped with randomly placed holes. They may constitute a smaller (holey film) or larger (lacey film) part of the film. They are a good choice if our nanomaterial has the form of flat or elongated particles. In such a situation, if they are located above the hole, they can be examined without an additional signal from the substrate, without losing a wide field of observation (which would occur if we choose a delicate metallic mesh, nontransparent to electrons). A certain variety of this group of substrates are Quantifoil films, with regularly arranged holes, known diameter and spacing. This type of meshes are particularly readily used for cryogenic microscopy (so-called cryoEM), as a support for micrometric amorphous ice droplets with a biomacromolecules or nanoparticles suspended inside.

– Metallic grids with a thin film of holey supporting film and additional layer – Sometimes, the holes from the previous group are additionally covered with an ultra-thin layer of the appropriate supporting substrate. This may be, for example, graphene, which sometimes offers a more favorable contrast than amorphous carbon. It also happens that the most delicate types of films can be mounted directly on a metallic mesh with the finest possible structures (mesh 2000 type).

– Silicon nitride (SiN) membrane, sometimes with regular holes – These types of substrates are mechanically the most durable and stable of all those mentioned above. They usually have a thickness ranging from 20 to 100 nm and therefore scatter the electron beam quite strongly and make high-resolution imaging difficult. Their high strength allows, for example, cell or bacterial cultures to be performed on them, and in the case of holes, it is a magnificent substrate for imaging 2D materials. Sometimes the dimensions of a single "window" are 0.5 × 0.5 mm (Figure 5.1c) or even 1.0 × 1.0 mm, providing a wide field of observation without covering the sample with metallic bars. However, high stiffness and durability has its price – these types of membranes are extremely brittle and should not be touched with tweezer or other tool under any circumstances.

The list mentioned above does not exhaust the topic. We can also mention metallic meshes with a single slot (commonly used for carrying the largest sections from an ultramicrotome), dedicated carrier chips for in situ TEM holders, and many others. I certainly encourage the reader to check the prices and availability of various types of substrates from a local supplier. In case of supply chain problems (as in some parts of the world during the 2019–2021 pandemic), it is also worth having a protocol in place for preparing some basic substrates yourself.

Once we have the solution of our nanomaterial and the appropriate substrate, it's time to start preparing the sample. Usually, it is enough to attach the grid to the cross-shaped tweezers (normally closed) or hold it with the tweezers on a neutral surface (which may be a glass or plastic Petri dish, aluminum foil, parafilm, etc.). Then we apply 2–5 µL of the sample using an automatic pipette, and leave the TEM sample until the solvent dries or excess liquid can be blotted beforehand. There is no single and unique rule – solutions with higher concentrations may require dilution or blotting, and samples with low concentrations may require large drops left to dry completely. I encourage you to prepare your grids in various ways to learn the practical aspects of achieving the perfect concentration of the sample on the substrate. It is also worth considering the use of anticapillary cross tweezers, which sometimes accidentally throw the sample into the air (when the sharp point pops off the opposite tip), but in return prevent the liquid from being pulled up through the gap near the sample. Of course, you should also take care to clean the tools after each preparation.

The preparation of 2D materials, which must be transferred to a TEM substrate, is much more demanding. In this case, the dominant and recognized method is wedging transfer, but there are also alternative and less proven, although not necessarily worse, solutions. These types of protocols require a steady hand and some experience, and access to a device that allows for CO_2 critical point drying (CPD) of samples will be beneficial. The ideal substrate for stable mounting of delicate 2D materials is SiN substrates with regular rows of holes, which allow imaging of the sample without additional contrast from the supporting film. You can also consider using metallic meshes with the smallest holes (mesh 2000).

5.2 Preparing TEM samples from bulk material

We mentioned at the very beginning of the chapter that one of the most important factors in describing a sample is its thickness. In the case of solid samples, the target thickness of 20–100 nm is achieved using various thinning methods.

The first method, which I will deliberately discuss separately, is ultramicrotomy. It comes directly from life sciences and involves the methodology of preparing thin sections from fixed fragments of biological material. It involves using a sharp glass or diamond knife to cut a sample embedded in a small epoxy block into slices of the desired thickness. This is an absolutely basic method, used, for example, to image biopsies, tissue fragments, and cell or bacterial cultures. However, it requires precise chemical or cryogenic fixation, followed by a relatively long dehydration procedure (replacing water with an organic solvent and epoxy resin), curing, and preliminary cutting for light microscopy (to assess which area will ultimately be examined on TEM). For precise descriptions of the methods of preparing tissue material, I refer readers to more specialized sources, but I would like to draw attention to the possibility of using ultramicrotomy methods in a slightly less conventional way. Namely, if your sample is relatively soft (polymer, multilayer composite, organic capsules, and carriers of drugs or nanomaterials), you can also embed it in an electron-resistant resin and then cut it into sections using an ultramicrotome. This will save you from having to use the multistep procedures described below and will be especially beneficial if your sample is less than 3 mm (standard TEM sample diameter).

A widely popular group of TEM samples are thin foils (Figure 5.1d and 5.1e), made from bulk material, usually metallic. To produce such sample, several steps need to be performed:

1. **Preliminarily cutting the material to dimensions that fit the grinding holder, and a thickness of approximately 0.5–1.0 mm.**

At this stage, we must try not to unnecessarily heat up the material or introduce unnecessary stress into it. This type of cutting should be performed using laboratory cutting machines with a coolant supply and controlled (small) pressure of the cutting disc on the sample.

2. **Pregrinding to a thickness of approximately 100 μm and polishing the sample in an adjustable grinding holder**

At this stage, we attach our material to the grinding holder using wax/thermoplastic glue (the author prefers this method) or contact glue (here the sample is removed with acetone, but there are problems with its disassembly). In the case of wax/thermoplastic glue, we should heat the holder on the heating table, apply a small amount of wax and our sample, and then cool the whole thing with a cool water stream. The holder allows you to gradually move the sample toward grinding paper during grinding (Figure 5.2). Ideally, the sample should be sanded alternately on both sides, changing sides approximately every 100 μm in thickness. This requires repeated heating and cooling of the holder for melting the wax, but allows for the minimization of stresses inside the material. If, out of haste or laziness, we grind the sample asymmetrically, there is a fairly high chance that it will be deformed under the influence of stress to such an extent that it will be over 100 μm thick in some places and perforation will occur in others. As we approach the thickness of 100 μm (controlled with a caliper, micrometer, or dial gauge), we finish the grinding process with polishing on diamond suspensions until a mirror shine is achieved (author prefers 5 μm, then 1 μm diamond paste on separate polishing discs). It is not required in every case, but it generally helps to obtain repeatable, high-quality results.

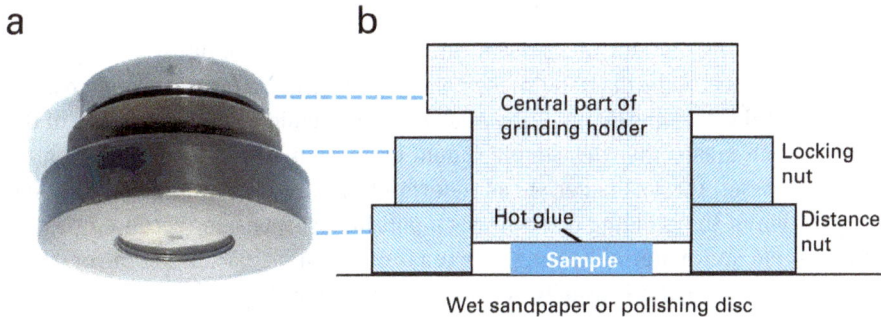

Figure 5.2: Laboratory-made holder for preliminary thinning samples for TEM: (a) image of the holder and (b) cross-section schematic of the device with sample and sandpaper.

In a situation where our priority is not to introduce any stress into the material (e.g., if we want to observe the dislocation structure of the material), it is worth considering replacing mechanical polishing with preliminary electrolytic polishing. Some manufacturers even supply separate holders with polishing machines for the initial polishing of larger samples and the final polishing of samples with a diameter of 3 mm. This type of approach requires some experience and attention (because we do not want to perforate the sample), but it is worth remembering this possibility.

3. Cutting discs with a diameter of 3 mm

When our plate is about 100 μm thick and has a mirror shine, we need to cut out discs of a size appropriate for TEM. The most convenient devices for this purpose are punching machines with a tool steel blade, but if our sample is exceptionally hard, an electro-discharge cutting machine can be used. In this case, we must equip it with an electrode in the form of a tube with an internal diameter of 3 mm. Another alternative is the use of ultrasonic cutting machines, which are particularly useful when dealing with nonconductive and/or brittle materials.

4. Dimple grinding involves thinning the central part of the sample to approximately 5 to 10 μm (optional)

This method is particularly applicable to multicomponent and multiphase metal/metal or ceramic/metal composites as well as materials that do not conduct electricity. Using a "dimple grinder" device, we make a deepening in the form of a sphere part on both sides of the sample, which is achieved by combining the rotation of the table with the glued sample and the rotation of the grinding wheel located above, reinforced with a diamond paste (Figure 5.3a). The geometry and process parameters should be adjusted so that the final thickness in the center of the sample is within the range of 5–20 μm, and in the last step, the sample should be polished to a mirror-like finish using the finest available polishing agents. A dial gauge mounted on the device can help track the amount of material removed (Figure 5.3a), but a more precise method is to measure the diameter of the polished area using a measuring microscope and calculate the grinding depth (5.3b). At all cost you should avoid: perforation of the sample (which means that the sample should be thrown away), its deformation (at small thicknesses, the soft material can deform plastically instead of being cut), and dry operation of the grinding disc without a polishing suspension (because of strain in the sample and material heating). Fortunately, in the situation of conventional metal alloy thin foils, this is usually not a necessary preparation step.

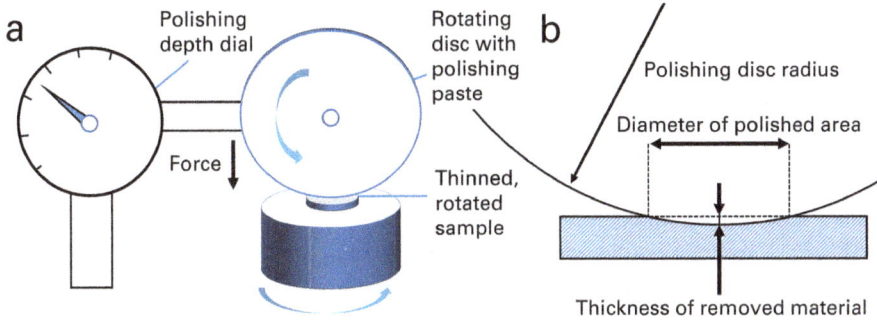

Figure 5.3: Dimple grinder for thinning central part of the solid sample: (a) scheme of dimple grinder operation and (b) geometry of polished area.

5. Electropolishing

This is by far the most popular method for the final thinning of metallic samples for TEM. It assumes that even polishing on both sides of the sample grabbed by the edges leads at some point to perforation, which, thanks to the use of a system with a photo-detector and a light source, leads to stopping of the TEM power supply and therefore the polishing process (Figure 5.4). On the border of the resulting perforation, there are areas several to several dozen nanometers thick, which offer a relatively wide area for observations. Moreover, in the electron-transparent area we will find areas of both smaller and larger thickness, which is extremely beneficial when we plan to use various imaging methods.

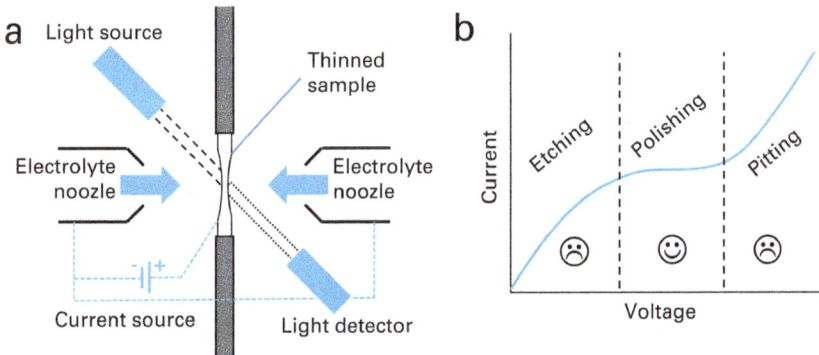

Figure 5.4: Electropolishing TEM samples: (a) scheme of the basic electropolishing device and (b) simplified shape of current-voltage function with marked electropolishing region.

Electropolishing is a process in which many variables must be controlled. These include the chemical composition of the electrolyte (many of them contain perchloric acid in an organic solvent, but depending on the material prepared, many different mixtures can be found), its temperature (room temperature or very often reduced temperature), the intensity of the electrolyte flow (ensuring its continuous exchange, but not causing turbulent flow near the sample), and current-voltage parameters. Usually, you should remember the shape of the current-voltage curve (Figure 5.4b) and work on its flat area (plateau). If there is a problem with selecting the initial parameters for popular alloys and reagents, it is worth trying current values in the range of 50-100 mA (with appropriately adjusted voltage). What is very important is that immediately after interrupting the polishing process due to perforation, the sample holder and the sample itself should be rinsed several times in a clean solvent (e.g., ethanol or methanol) to inhibit the processes of surface corrosion and oxidation. In the case of many materials and research protocols, the electropolished sample is suitable for TEM observations. However, it may turn out that its surface is locally covered with oxides or other process residues, which manifest themselves in the image as amorphous areas. In this case, the simplest method to improve the sample is short (5–15 min for wide-beam devices like GATAN DuoMill, and 3–5 min for more effective devices like GATAN PIPS) ion polishing. In the case of multiphase materials with complex microstructure, it may turn out that some components of the structure are polished much faster than others and not every phase is observed in TEM as transparent. In such a case, you can experiment with higher current values, perform preliminary dimple grinding, or use additional, final ion polishing for longer time.

6. Ionic polishing (optional)

This preparation method can be used to clean and refine the sample after electrolytic polishing, to remove excessive contamination after long-term observations, or as an independent preparation method. It is particularly important for samples that do not conduct electricity and cannot be subjected to electrolytic polishing.

Ion polishing is actually abrasive blasting in which argon ions are our cutting medium. They are accelerated with high voltage, and during the process you can observe plasma emerging from the ion gun. The most common ion polisher setup for TEM involves a rotating holder with a mounted sample with both sides (top and bottom) exposed to plasma (Figure 5.5). On the sides of the sample there are two ion guns facing each other, mounted at a small, adjustable angle to the sample plane (from as much as 1° in some devices to over 20° in others). The controlled parameters include argon flow, process angle, and voltage applied to the guns. It is worth remembering that larger angles generally intensify the polishing process, but make it difficult to obtain a wide viewing area. In turn, the smallest angles sometimes encounter a situation in which the geometry of the device, holder or sample itself means that the ion guns do not "see" the hole and do not polish its area. Although ion polishing is one of the most delicate preparation methods, too intense polishing may lead to local heating and degradation of the sample (for this reason, in some devices, it is possible to cool the sample with liquid nitrogen). Another disadvantage of the discussed process is its low speed. For many devices, a realistic and reasonable limit is to thin the sample by ap-

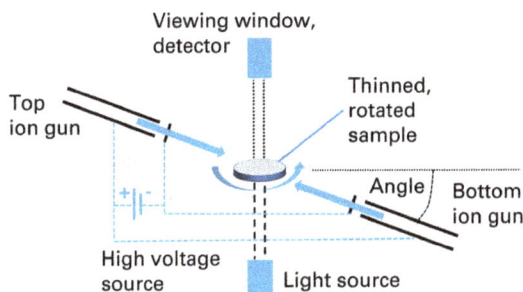

Figure 5.5: Scheme of ion polisher for TEM sample preparation. Keep in mind that whole setup remains in high vacuum.

proximately 1–2μm/h, with ion guns requiring complex maintenance often after less than 100 h of operation. For this reason, if ion polishing is our main method of final thinning, it should be preceded by preliminary grinding using a dimple grinder.

5.3 Preparing TEM samples using a focused ion beam (FIB)

One of the most demanding preparation methods is requiring thin layers applied to the substrate. In conventional preparation, we use the so-called sandwich technique, i.e., two small, precisely cut fragments of the sample are glued in layers to each other using an electron beam-resistant glue. Then we glue the whole thing inside a metallic tube with an outer diameter of 3 mm and fill the free spaces with appropriately matched and cut fragments of a rod that fits the tube. It is very important that the fit is not too tight (the sample will fall out during preparation because of internal stress) or too loose (large areas of glue are more likely to perforate than a substrate or layer with a higher density). After the glue has cure, the whole thing is cut into thin slices with a diameter of 3 mm, which are then treated as a solid material (pregrinding, then dimple grinding, and ion polishing). However, practice shows that the method described above has a relatively low success rate. On this occasion, I would like to describe an alternative method for the preparation of thin layers, which can also be used to prepare almost any material, the FIB method.

The described method is usually performed in dedicated scanning electron microscopes (SEM), which are equipped with an additional ion gun (FIB). The entire system allows for local material removal as well as targeted deposition of various materials (such as carbon or platinum) from precursors introduced into the microscope vacuum. In practice, this means that we can locally cut out a fragment of the material from (usually) its surface layer, extract it using a micromanipulator, and obtain an electron-transparent sample (the so-called lamella), with dimensions from 5 × 5 μm to about 20 × 20 μm. It is mounted on dedicated, half-moon metallic substrates (Figure 5.1e) and finally thinned to a thickness appropriate for TEM. This method is particularly useful for analyzing the structures of surface layers, but is also often used as an alternative to all the methods described above. Its advantages include universality, the possibility of installing lamellas on dedicated chips for in situ TEM, as well as the possibility of precisely cutting the lamellas from specific sample region (so we can, e.g., check the structure within the place of interest at the fractured surface or a specific place in a complex semiconductor system). The disadvantages include the relatively small size of the sample, which are often several dozen or several hundred times smaller than the transparent area of thin foil (Figure 5.6). This makes it difficult to look for specific crystallographic orientations, specific defects, or grain boundaries. There is no doubt, however, that nowadays FIB-SEM preparation is a strong, interesting, and useful alternative to conventional preparation methods. It is definitely worth having it in your range of available techniques.

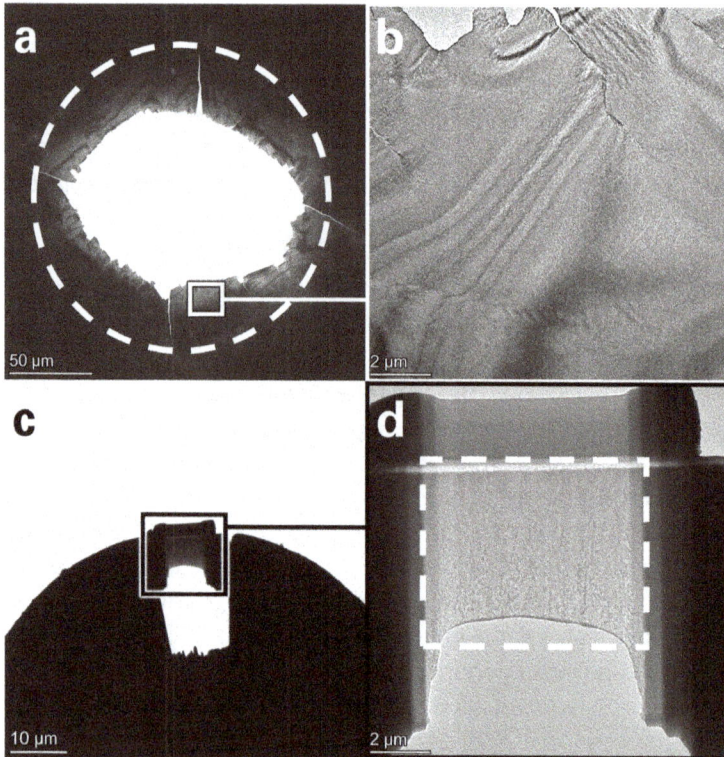

Figure 5.6: Comparison of transparent area size in metallic thin foil (a, b) and FIB lamella (c, d). Electron transparent areas marked with dashed line. You can notice that thin foil offers hundreds times more observational area, what could be beneficial.

5.4 Additional comments and notes

A trivial but important piece of advice is to carefully read the operating instructions for the devices you use. Often, among the basic facts, you can find really valuable advice that will allow you to optimize the preparation process or explain the most frequently occurring problems and artifacts. Some devices also have special, dedicated procedures for setting certain parameters (such as electrolyte flow in an electrolytic polisher), which greatly simplifies achieving good results.

As an additional tip, I suggest that you carefully describe the preparation history of your samples in your lab book. This not only allows you to refine the process parameters for the new material but also facilitates work in the future. Refining the preparation method can be tedious and time-consuming, which is why it is worth keeping records of proven protocols for the types of materials already tested for your future needs.

Sometimes, it is also worth spending one type of the sample and some time to observe the impact of changing preparation parameters on the width and shape of the transparent area. Simple experiments to perform include, for example, repeated ion polishing for 15 min with observations of the same sample between subsequent processes, or a comparison of the effects obtained for the same polishing time and different ion polishing angles.

Regardless of the presented information, I encourage you to consult more detailed literature sources [6–9].

Take-home messages

- When preparing your TEM samples, remember that they must meet several criteria. First of all, they must be locally transparent to electrons, stable in high vacuum conditions, relatively resistant to electrons, representative of the tested material, and not modified by the preparation method.
- For nanomaterial preparation, a good starting point is a simple carbon film on a metallic mesh grid, but consider using variety of different supporting grids.
- For metals and alloys, TEM sample preparation includes mechanical cutting and prethinning, grinding, and polishing to a thickness of approximately 100 µm, electrolytic polishing, and sometimes finishing ion polishing.
- For nonconductive or multiphase materials, dimple grinding and ion polishing can be used instead of electropolishing.
- It is worth describing the preparation of your samples in detail in your lab book because, due to the many available parameters, ideal protocols for different materials may require extensive optimization.

Literature

[1] Williams DB, Carter CB. Transmission electron microscopy: A textbook for materials science. 2nd ed. New York: Springer; 2008.

[2] Cosslett VE. Radiation damage in the high resolution electron microscopy of biological materials: A review. J. Microsc. 1978; 113(2):113–29.

[3] Jenkins ML, Kirk MA. Characterisation of radiation damage by transmission electron Microscopy. IOP Publishing Ltd; 2001.

[4] Egerton RF. Control of radiation damage in the TEM. Ultramicroscopy 2013; 127:100–8.

[5] Egerton RF. Radiation damage to organic and inorganic specimens in the TEM. Micron. 2019; 119:72–87.

[6] Echlin P. Handbook of sample preparation for scanning electron microscopy and X-ray microanalysis. Boston, MA: Springer USA; 2009.

[7] Ayache J, Beaunier L, Boumendil J, Ehret G, Laub D. Sample preparation handbook for transmission electron microscopy methodology. New York, NY: Springer New York; 2010.

[8] Ayache J, Beaunier L, Boumendil J, Ehret G, Laub D. Sample preparation handbook for transmission electron microscopy techniques. New York, NY: Springer New York; 2010.

[9] Kuo J. Electron microscopy: Methods and protocols. 3rd ed. Totowa, NJ: Humana Press; 2014. (Methods in molecular biology; vol 1117).

6 Using a microscope – basic observations

In this chapter, I will guide you through a universal protocol for basic TEM imaging. It should work on most of the equipments available on the market (or still in use in different laboratories), but it is possible that in some exceptional situations, you will encounter significant differences in work principles. The basic and most important rule is to obtain information directly from your supervisor, laboratory manager, local technician, or a more experienced colleague. It's better to ask three times unnecessarily than to turn the wrong knob once.

In modern designs, the operator panel is kept to a minimum. In older ones, you can find several dozen keys and knobs [1, 2]. Nevertheless, you can find knobs or buttons described as

- Condenser/intensity/beam focus – they control the current flowing through one of the lenses of the condenser system (usually the second, C2, but in the more complicated TEM/STEM systems, this does not have to be the rule).
- Beam shift/condenser shift/beam alignment – these move the beam within the area of interest and allow the beam to be centered on the screen or the camera. They are often available as the basic function of "multifunctional knobs," which, if necessary, can adopt one of several dozen functionalities.
- Magnification – by appropriately changing the current in the intermediate/projector lenses (but not only!), it allows you to obtain various magnifications during the imaging.
- Focus – moves the focal plane up and down to obtain the appropriate focus value.
- Stage/XY/sample – They respond to the movement of the sample inside the microscope to select the area of interest. Often located nearby are the Z-position control (although sometimes it may be in the form of a physical adjustment knob on the goniometer) and the sample tilt angle control. In the past, the latter functionality was often connected to control pedals, which allowed for simultaneous tilting of the sample and correcting its *XY* position using hand-operated controllers.

To work freely and effectively, we also need access to a few additional parameters or functionalities. They may be hidden in the software, located on the operator panel, or found in some less frequently used panels. These include:

- Reset/eucentric focus/zero focus button – it runs a command that resets the current flowing in the objective lens (and sometimes in all lenses) to the factory value at which the microscope should maintain the highest performance. In some designs (such as JEOL microscopes from 1970to 2000), this functionality is implemented by displaying the values of currents flowing through the lenses on the information screen and setting the "focus" value in such a way that the objective

https://doi.org/10.1515/9783111317014-006

lens (described as OBJ) has some specific current, described by user manual (and is specific for different devices).

– Wobbler/beam wobbler/focus wobbler – this command starts a mode in which the condenser system starts illuminating the sample alternately from two different angles. This causes the image to "split" in a defocus situation, and as the Z-position or focus position changes, the image becomes still or almost still close to zero-focus position.

– Objective wobbler/OBJ wobbler/HV wobbler/rotation center – This command, in simple words, allows us to align the electron beam parallel to the optical axis of the objective lens. In some systems you can find separate "OBJ wobbler" and "HV wobbler" settings, and sometimes only one of them. There is no point in setting both parameters alternately. If possible, a perfectly set "OBJ wobbler" may cause minimal "HV wobbler" misalignment and vice versa – without affecting the quality of observations. In such a case, a simple rule may be consistency and adapting to the habits of other users. Basically, both commands perform a quite similar task of setting the beam tilt to "zero" when observing a specific sample. This is important and visible especially when imaging ferromagnetic samples, where the correct setting of this parameter, may differ between samples. A clear indication of the need for calibration with an OBJ/HV wobbler is a situation in which the sample image clearly moves in the detector plane when moving the focus plane. The exact implementation of this calibration will be discussed later.

– Beam tilt/rotation center/bright tilt – This pair of knobs (sometimes one of the multifunctional knob modes) is responsible for adjusting the tilt of the beam hitting the sample. It is used to calibrate the sample in the case of the OBJ/HV wobbler calibration described earlier (and as a "dark tilt" for dark field observations).

– SA/DIFF or TEM/DIFF switch for switching between imaging and diffraction modes – sometimes in the form of a bistable switch or buttons. Changes the operating mode of the projector system in a way that allows the observation of the diffraction plane (diffraction obtained as a result of the interaction of the beam and the sample) or the image plane (the magnified image of the sample in a given focal plane)

– STIG/stigmators – knobs controlling the condenser, objective, and diffraction stigmators. They allow for the necessary correction of astigmatism and are sometimes hidden as multifunctional knobs mode or hidden in the auxiliary, side panels.

– Apertures – selection of specific aperture sizes (condenser, objective, and diffraction) and their positioning mechanism. Sometimes these are physical knobs on the column, other times they are commands implemented on the operator panel or in software that translates into the movement of electric step motors.

Now that we have this set of basic tools figured out, we can move on to the basic electron observation protocol. Please do not treat it as an oracle or the only correct way of working with a microscope, but as a collection of tips and basic information before the actual training with the microscope supervisor or laboratory manager. I have tried to describe the procedure from the perspective of the user of most systems available on the market, but it may turn out that in the case of your device, some of the procedures look different or are even absent (e.g., Section 6.2 is not possible to perform). The device has many more parameters, but for the purposes of this simple introduction, we will skip some of them. You will definitely start using them yourself as you gain more experience and consult with more experienced users.

6.1 Starting TEM

Before you start working, you need to make sure that your device is in good condition. In the case of microscopes with thermal tungsten emission, it may turn out that you need to "cold start" the microscope, which will be operational after 30–60 min. In the case of laboratories that are able to ensure uninterrupted power supply, such a microscope can be turned on 24/7, and if we are dealing with a LaB$_6$ or field emission gun, the microscope will most likely already be running, so you can start working immediately. In the operating manual you will find the vacuum values that should characterize the system ready for operation as well as additional advice on starting or controlling the gun. If your microscope is equipped with an anticontamination device (ACD), it may also be time to fill it with liquid nitrogen. However, sometimes users prefer to perform this step after installing the sample and seeing the image. This is due to the fact that for some maintenance activities (such as replacing a burnt emitter) you need to have the ACD in a warm state [3].

It is also worth paying attention to whether the device is in its "basic" state— whether the lenses are activated (sometimes they require pressing a specific button or switch) and whether there is no unnecessary aperture inserted or an electron camera blocking the fluorescent screen. It is worth trying to prepare a "checklist" for the beginning and end of work, which will ensure that each user leaves the microscope in the same and appropriate condition before starting work the next day.

6.2 Obtaining a beam image without a sample

One of the most common problems for TEM operators is the situation in which, after installing the sample and starting observations, we observe only darkness on the fluorescent screen or camera. The only certainty in such a situation is the fact that the electron beam does not reach the detector. However, we do not know what is blocking it. Maybe the aperture is unnecessarily inserted or incorrectly centered? Maybe a system that deflects the electron beam illuminates a place outside the screen? Maybe the electron gun isn't emitting a beam? Maybe we have closed some valves in the column? Maybe we are simply observing a nontransparent part of the TEM sample or just grid bar?

To facilitate the diagnosis of such situations, it is worth making an image of the electron beam without the inserted sample in the first attempt. We perform basic beam alignment many times on a daily basis, during every typical job and with a sample inside. This time, however, we want to make sure that the beam is not, for example, excessively focused, excessively expanded, or moved outside the screen. Not every microscope will allow the generation of the beam without a holder inside – in such a situation, it is enough to mount the holder inside without the sample. Some systems (such as some Hitachi designs) have the additional option of placing the sample in a "parking position" – it is already in the vacuum and near the objective lens polepiece, but the holder is still completely outside the electron beam.

Regardless of the path we choose, we should provide basic conditions for observing the beam. This usually involves removing the objective and diffraction apertures, but leaving the condenser aperture centered (some systems do not even allow it to be fully removed). We should switch our microscope to the TEM observation mode (sometimes called TEM, other times ZOOM, SA, BF, or similar), at the lowest possible magnification (usually several thousand times). It is also worth resetting the lens settings so that the focus plane is in the optimal, factory spot. Then we start observations and start generating and previewing our beam.

ℹ Different electron emission type controls

This stage of work will differ significantly for microscopes with thermal, thermal-field Schottky, and cold field emission [4]:

– W or LaB$_6$ thermal emission may require gradual activation of the accelerating voltage (according to instruction manual) while observing the emission current and vacuum readings. You may find that contamination of the emitter will cause a temporary deterioration of the vacuum and will require some time to restore. After starting the target voltage (usually maximum), the cathode is heated (using a knob described as emission or filament heating). The emission value should be brought very gently to the appropriate value (the entire range of the knob to about a minute or slower). If the system is in continuous use, this knob position will likely be marked with a "stopper" limiting further movement. In the second half of the rotation you should notice a gradual increase in the emission value and the appearance of the beam image. In the case of LaB$_6$ microscopes, you may also encounter a situation where the beam will be blocked by a valve between the gun and the column – of course, you must ensure that it is open during operation.

– Schottky field emission is generally computer-controlled and emitters of this type usually operate 24/7 due to the finite life of the emitter in terms of the number of its on/off switches. In this case, you should make sure that the extractor voltage, emission current, gun, and column vacuum values are within the recommended instruction manual values, then open the valve between gun and column, releasing the electron beam.

– Cold field emission is also computer-controlled, but its start requires first carrying out an emitter cleaning procedure, the so-called flashing. Then, emission is started in a similar, digitally controlled way. This procedure varies between systems, so refer to your user manual. Here you also need to remember to open the valve between the gun and the column and monitor the drop-in emission current during operation (and repeat some type of flash if necessary).

Many systems are equipped with a valve that isolates the gun from the rest of the column. A good practice is to get into the habit of closing it whenever you are not performing observations. For example, if you need to leave the device for a moment, plan to replace the sample or simply finish a certain stage of work. This type of approach increases the probability that, in the event of an unexpected failure (power cut, cooling water cut, vacuum leak), the gun will remain in good condition as much as possible.

If we see our beam, it is worth ensuring that its position and shape make future observations easier. For this reason, we choose the lowest possible magnification and center the position of the beam. You need to focus it slightly with Intensity knob (if you are working on a fluorescent screen, you can even focus it to the size of a small spot), set it in the center of the screen with the Beam Shift knobs and then expand it so that it covers the entire screen (or even a little wider). If you see that the beam does not expand symmetrically and concentrically, see Section 6.3, and if the beam is not round but elliptical, see Section 6.4.

If the shape of the beam satisfies you, in the next step, you can close the valve or additionally turn off the emission (in systems with thermal emission) and prepare for the sample installation. However, if your sample is not placed on a supporting grid, but a material that is more difficult to orient (such as a thin foil in which you first need to find a hole with a transparent area), it is worth considering using the low magnification mode (Low Mag or LM), which will make it easier to observe a wide range of sample area. It is worth using the same mode if our sample is hard magnetic and your holder offers a relatively weak hold on the sample. In such a situation, the LM mode ensures minimal excitation of the objective lens, which prevents the sample from being torn out from the holder when entering the magnetic field of the activated objective lens (reaching up to 2 T!).

6.3 Inserting and centering the condenser aperture

During routine operation in TEM and SAED modes, it may turn out that the condenser aperture is practically not touched by the user for many months. However, any change in the condenser aperture (e.g., when changing between the STEM or TEM mode of operation, or if necessary to control the beam convergence angle) requires its centering. Symptoms of an off-center aperture position will include (but are not limited to) nonconcentric beam focusing (Figure 6.1). First, we need to determine from which side of the focus of the condenser lens we should work. In some designs, from the position of the focused beam (to the minimum spot), we should expand the beam clockwise, and in others, in the opposite direction. This is usually described in the user manual, or you can ask other users about it. In principle, using the opposite side of the focus is not a critical mistake, but if we care about the quality of our work and results, it is worth consistently using only one side of the condenser. If you can see the currents flowing in the lenses, we are heading toward underfocus, i.e., decreasing lens currents as we expand the beam with the Intensity knob. Usually, the second condenser lens (C2, K2) is responsible for this function, but this does not always have to be the rule [5].

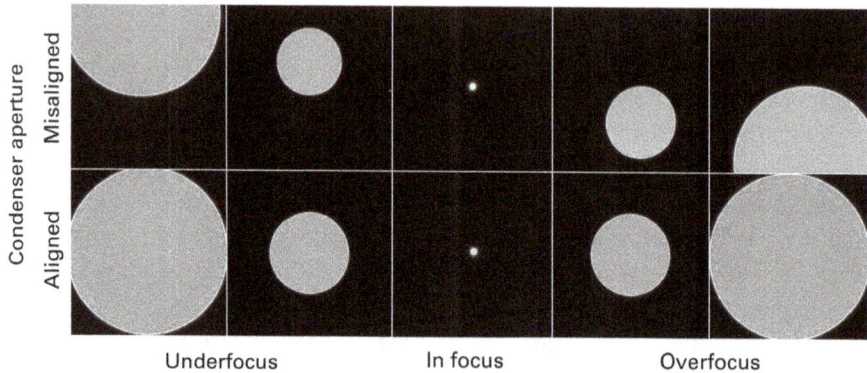

Figure 6.1: Comparison of beam images for misaligned and aligned condenser aperture during focusing the beam using the Intensity knob.

If we want to fix a nonconcentric beam image, we should focus the beam to a small point with Intensity knob (Figure 6.2a), then use the Beam Shift knobs (deflectors) to center it to the screen (Figure 6.2b), then expand it to the underfocus position (Figure 6.2c) using the Intensity knob (but only a bit, so that we can clearly see the position of the expanded beam in relation to the screen), and then center the expanded beam by aperture shift (using manual knobs or using the operator panel or

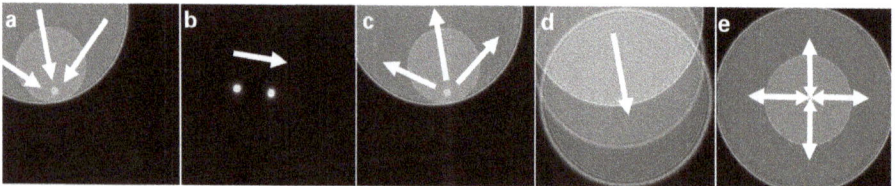

Figure 6.2: Procedure for centering the condenser aperture centering: (a) focus the nonconcentric beam using Intensity knob, (b) center the spot using the Beam Shift knobs, (c) expand the beam using the Intensity knob, (d) center the wide beam disc using the aperture shift, and (e) focus and expand the beam using Intensity knob and repeat in necessary.

software) (Figure 6.2d). Then we focus the beam again, center it using Beam Shift, expand (Figure 6.2e) and center it using aperture shift. After two to four iterations, the condenser should behave repeatably and correctly. Importantly, if we move to a position on the other side of the focus in a perfectly centered condenser, there may be some deviation from the ideal concentricity of the beam. So if you have observed this phenomenon for the first time in a long time, first check whether you have accidentally moved to the other side of the condenser focus (to the overfocus position).

If the same symptom of nonconcentric expansion of the condenser occurs with the condenser aperture pulled out, the position of the condenser lenses should be centered, which is usually quite a complex service operation.

6.4 Condenser astigmatism correction

If our condenser is concentric (or even not), but we notice that it does not form a regular circle, but an ellipse inverted at an angle of 90° after passing through the focus – this is a sign that we are dealing with condenser astigmatism (Figure 6.3).

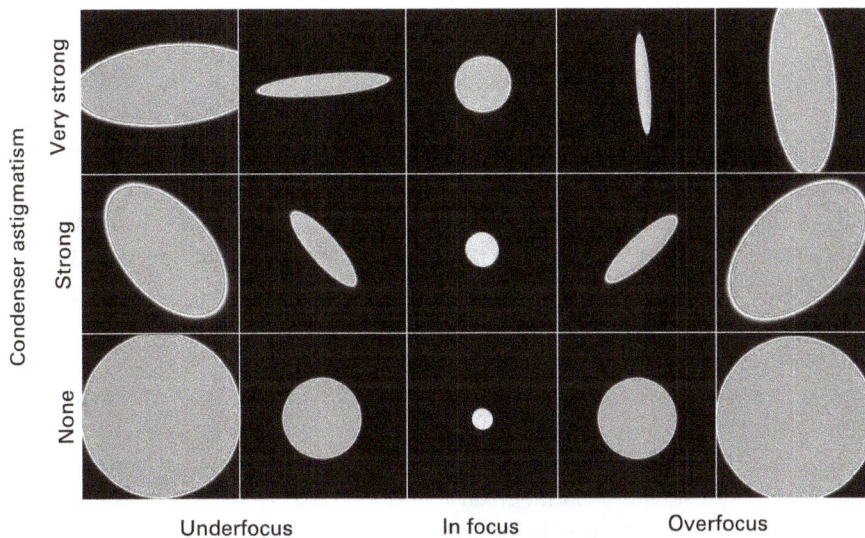

Figure 6.3: Comparison of astigmatic and nonastigmatic beam shape.

Adjustment of condenser astigmatism is a bit easier than centering the aperture. We must ensure that the beam is centered and fully visible within the field of view. Then we use the Cond Stig knobs (or a similar function, which is also available as one of the multifunctional knobs functions) and use them to correct the circularity of the beam. My advice is to adjust the image using each knob alternately rather than turning both knobs at once. In such a situation, we first correct our beam to the least elliptical position and then correct the shape with the second knob. Two or three repetitions should be sufficient. You will probably notice that as you get closer to the condenser focus (focusing the beam), the astigmatism may change. For this reason, it is worth correcting it at relatively high magnifications and with a focused beam because it is in such observation conditions that we care most about the quality and the shape of the beam. An alternative method of correcting condenser astigmatism is to use higher magnifications, concentrate the spot to a point as much as possible, and set the stigmator so that the spot has a minimal and round shape. After expanding the beam, it may turn out that we are dealing with a slight ellipticity of the wide beam, but in observations at low magnifications its shape is less important. In principle, it is often possible to adjust the astigmatism in such a way that both the narrow and focused spots are round within our perception. If the astigmatism cannot be corrected (you can see that you can influence the shape of the beam, but to an insufficient extent), you will probably need to clean or replace the condenser aperture. A temporary solution may be to choose a different size of the moving aperture, center it, readjust the astigmatism, and continue the observational work with a slightly different intensity and beam convergence than the previous try.

6.5 Preparing for sample installation

Once we are sure that our condenser system works perfectly and that nothing is blocking the beam on its way to the sample and detector, we can confidently install our sample. If your laboratory's habits are different, you can also start working with sample installation. Due to the fact that the moment of replacing the specimen holder is generally the most risky moment of working with TEM, it is worth remembering a few basic issues:

–   vacuum values in the gun and column must be sufficient and correct,
–   for thermonic electron gun it is worth turning off the emission and/or closing the gun valve; for field emission gun, it is necessary to close the gun valve,
–   the position of the goniometer (and/or the sample holder inside, if you have not removed it yet) should be in the neutral position (usually with all values: X, Y, Z, α, and β close to zero). This is important because in some microscopes and/or holders a collision may occur during removal, if the values are inappropriate,
–   objective aperture as well as additional and delicate detectors (such as the slide-in EDS) should be removed.

6.6 Installation of the sample into the sample holder

Sample holders from different manufacturers and types differ significantly in terms of mounting the sample. You can find holders with a movable flap on a hinge, a cover attached with screws, or even with a single nut with an internal hole for the sample. In each of these situations, I suggest consulting the user manual or contacting an experienced user. In addition, most holders are equipped with a vacuum O-ring, which functionally divides the holder into two parts – ultimately located inside the column (closer to the sample) and located in the goniometer lock and the external atmosphere (closer to the gripping part). While the entire holder must be treated gently and preferably handled with laboratory gloves, the internal part should not be touched at all to avoid unnecessary contamination of the system.

It is worth paying attention to whether the sample should be installed with the supporting film facing up or down. Many systems require you to rotate the holder by approximately 120° to 180° when installing the holder into the column, so if you want a specific sample orientation, it is worth considering this. This is especially important for thicker substrates such as SiN mounted on massive Si. In such a case, it may turn out that improper positioning of the sample in the holder prevents the Z-axis from being correctly set without colliding with the internal elements of the microscope (or right Z-position is out of available range).

If you work with thin foils that have a hole, it is worth placing the hole as close to the main axis of the holder as possible. This will make it easier to find a transparent area and make subsequent diffraction tests (areas outside the main axis of the holder change their position more strongly in the X, Y, and Z-axes during tilting).

If your workstation is equipped with a low-magnification light microscope, binoculars, or magnifying glass, it is worth checking whether there are any unnecessary contaminants (dust, fibers, hair) around the sample. In the same way, it is worth inspecting the O-ring of the holder and, if necessary, removing impurities and/or gently lubricating the seal with an appropriate vacuum grease (after consulting the device manager).

It may turn out that after installing a dozen or so samples, they start to stick to the holder elements. In such a situation, it is worth using a dust-free wipe or a dust-free optical cleaning stick and a neutral solvent (isopropanol, ethanol, methanol) to gently clean the sample cavity and mounting elements. Then, it is advisable to leave the holder to dry without using additional heating devices. At the most, you can use a lamp with a tungsten or halogen bulb that will gently warm the cleaned areas. It is worth remembering that some solvents (e.g., acetone) have a harmful effect on rubber seals or plastic elements and should be used with great caution.

6.7 Installing the sample holder into the TEM column

In most popular systems, the installation of the holder in the column is done in a similar way. First, place the handle in a position allowing for pumping out the prechamber (pumping starts automatically or after some specific user action like hitting the switch). After obtaining the appropriate vacuum (measured in pumping time or with a vacuum gauge), the user is allowed to gently turn the handle (which simultaneously opens the valve between the intermediate chamber and column) and guides the holder inside, gently counteracting the pulling force of the vacuum. The holder must not be allowed to fall freely inside, which may result in a costly failure. It is also worth not making any unnecessary hand movements up-down or right-left and also not pulling back the holder, as this may cause leakage in the seals. On the other hand, many sample installation failures (i.e., excessive vacuum deterioration due to installation) among novice users are caused by their excessive gentleness and caution. However, there is no doubt that practice makes perfect.

If, after installing the holder, the vacuum has deteriorated significantly (i.e., installation caused a temporary leak), it is worth monitoring the readings of the vacuum gauges. If they return to normal quickly – it is perfect, if after a while – it is still good. However, if after a long time the vacuum is still not behaving normally, you may consider returning the holder to an intermediate position and checking if the vacuum does not improve quickly with the valve between the airlock and the column closed. If the system behaves this way, there is a chance that the leak is caused by minor contamination of the O-ring on the holder. Remove the holder, clean and lubricate the seal, and then try to install the holder again in the column.

6.8 Starting the observations and setting eucentric focus

If our sample is inside the TEM, and initial observations have shown that our beam is ready to go and in the right place, we are in the perfect position to start our observations.

So, open the gun valve and start electron beam emission. If you are working with the sample on a supporting film, you should be able to see the image quite easily (we made sure the beam was in the right place in Section 6.2 of our instructions). If you are working with metallic thin foil, it may be more difficult, but if the sample hole was in the middle you should be able to find it with low magnification. If you have placed the hole along the main axis of the holder, it is worth moving along this axis (which is sometimes called both the X and Y axis). In the most difficult variant, you have to "scan" the sample using a XY controls until you find a transparent place. If you make sure the beam is in the right place in step 6.2, all you need to do is to remain patient. Fortunately, when approaching the hole, you can notice a slight glow of deflected electrons hitting your screen or detector. If you are working on a fluorescent screen, make sure the room is adequately darkened – this will help you seeing even slight changes in the screen's light intensity.

If you have already found a place suitable for observations, it is worth making one or two basic adjustments to your TEM.

The first step is to set the appropriate position of the sample along the Z axis. We already mentioned at the beginning of this chapter that our lens has one position in which it works optimally (for the purposes of this description, we will call it the optimal focus position). For this reason, we must bring our microscope to a position where the base defocus of the objective lens is zero. This position is sometimes also called the eucentric position because in the ideal and most optimal situation (which, however, is not always fully true) it is also the place where the smallest center deviation along the XY axes occurs when tilting the holder along its main axis (tilt along the α axis). Therefore, we will call it eucentric focus position. As we have already mentioned, in an ideal situation, optimal focus position = eucentric focus position. Very often, however, there are some discrepancies between these positions (which are completely typical and related to the mechanical positioning of the goniometer in the column), and aligning the goniometer position is an important service operation after each disassembly. Below I will describe the method of positioning the sample in both positions (or in the same position with two methods, for the ideal situation):

- Optimum focus position – we need to make sure that our defocus = 0, or the current flowing in the objective lens is the base value (we discussed this when describing the operation of the reset light/button at the beginning of this chapter). Then, using the Z axis movement, we place the sample in the in-focus position (in the position of the lowest image contrast). It is worth using the previously discussed Wobbler function, which causes the image to move by tilting the beam in both directions. We then have to move the Z axis in such a way as to reduce the movement in the center of the frame to a minimum, as shown in Figure 6.4 (it can pulsate or rotate, but cannot move sideways). After completing the setting, turn off the Wobbler mode.

- Eucentric focus position – in this case, our goal will be to strive for a situation in which the sample image moves as little as possible when tilting the holder. Some systems are equipped with the α-wobbler function, which automatically tilts the holder and allows the user to focus on operating the Z axis in a way to minimize the movement of the central area of the observable sample. If our microscope does not have such a function (or we like manual solutions), we must (1) find a recognizable structure element at low (1,000–10,000×) or medium (20,000–100,000×) magnifications, (2) rotate the holder by several or a dozen degrees to one of the sides, and then (3) use the Z axis to bring the same element to the center of the screen. Then repeat the procedure with the opposite tilt direction and refine the calibration several times. However, please remember that the further we are from the main axis of the holder, the greater will be the movement of the sample, which cannot be fully corrected.

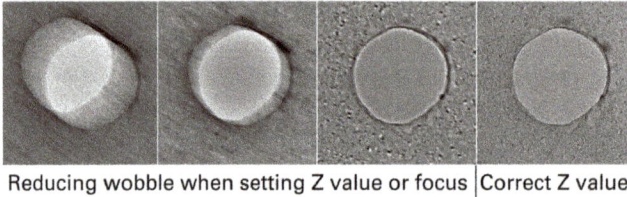

Reducing wobble when setting Z value or focus |Correct Z value

Figure 6.4: Typical images during setting the Z value or focusing, when using Wobbler.

The first method is primarily aimed at ensuring optimal operating conditions of the optical system and maximizing resolution. The second one is valuable if we plan to conduct diffraction studies that require frequent tilting of the sample (or tomography, but there the calibration procedure is even more demanding). For beginners and/or those who do not tilt samples a lot, I definitely recommend the first method as it is quick and effective.

The second important calibration (especially when observing ferromagnetic samples) is setting the correct rotation center (discussed in more detail in the description of the knobs at the beginning of the chapter). The symptom that you need to make this correction during work is a situation in which, when you turn the focus knob hard, you can see the microstructure elements clearly moving in a direction other than rotation or pulsation relative to the center of the screen. In such a situation, we use one of the functions called rotation center/beam tilt/HV wobbler/OBJ wobbler/BF tilt and set the Beam Tilt knobs (or appropriately set multifunctional knobs) so that the image rotates or pulsates relative to the center of the frame (Figure 6.5). Importantly, we should

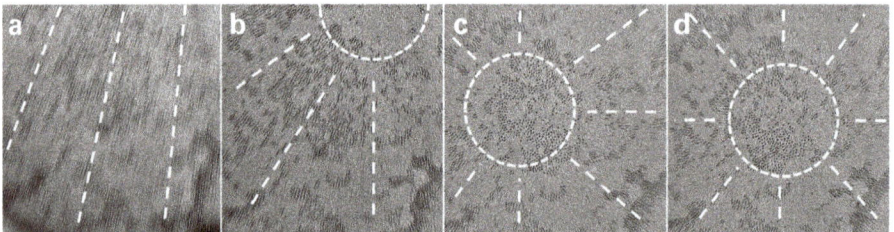

Figure 6.5: Setting the correct rotation center on quantum dots sample, dashed lines show direction of movement: (a) rotation center outside the field of view, (b) whole sample image is wobbling, (c) rotation center in the field of view, but sample image is pulsating not perfectly in the center, and (d) correctly adjusted rotation center, central part of the image remains stable when outside parts of image pulsate concentrically.

not look at the shape of the beam (let's center it with the Beam Shift and/or expand it so that it does not disturb us), but at the element of the structure of our sample in the center of the image field. It is very important to perform this correction after setting the Z axis and near zero defocus. After making the correction, we return to the observations. This setting is sensitive to the magnetic properties of the sample and must be performed each time, for example, when testing ferromagnetic thin foils, or when alternately testing samples of this type and more magnetically neutral substances.

6.9 Changing the magnification and the area of interests

We have already selected the place for observation, and the initial setting of the microscope has been made. Therefore, there is nothing left to do but conduct proper observations. At this point, our objective aperture is outside the beam, so we make observations in the so-called bright field (BF). This means that our diffractive central reflection takes part in the formation of the image. We will discuss conducting bright- and dark-field observations in connection with diffraction in Chapter 7, but now it is worth getting used to the fact that most observations are naturally conducted in the BF mode. Poorly transparent places will appear as dark (nontransparent as black), and completely transparent places (like a hole in a thin foil or supporting film) will appear as white. Diffraction phenomena are also involved in the formation of different shades of gray (sometimes even black), independent of the sample thickness, but the general rule remains the same.

The selection of different places for observation and their precise location in the frame is achieved by moving the sample along the XY axes (not by moving the beam), and the change in magnification is caused by a predefined change in the currents flowing through the lenses of the intermediate and projector systems. However, a very important skill that every TEM operator must acquire is the independent operation of the condenser and projector systems. If we choose higher and higher magnifications during our work (Figure 6.6a–6.6c), we will quickly notice that the image becomes dark and unreadable (and in the case of CCD/CMOS sensors, more noisy). This is because our condenser array still illuminates a wide area, typical for lower magnifications. For this reason, each time you increase the magnification, you need to focus the beam a little and make sure it is centered in relation to the screen. For this reason, you should use the Intensity knob to focus the beam so that its edges are visible, center it with the Beam Shift knobs (Figure 6.6d), and then expand it (Figure 6.6e) so that the beam covers the entire frame (or even wider, if we want to reduce the electron dose rate). If you notice that the beam is nonconcentric or elliptical, you should return to setting the position of the condenser diaphragm (Section 6.3 of the manual) or setting the condenser astigmatism (Section 6.4). If you reduce the magnification, you should expand and possibly center the beam so that its edges do not obscure the frame. It may turn out that your image will be significantly less sharp or distorted at the edges of the

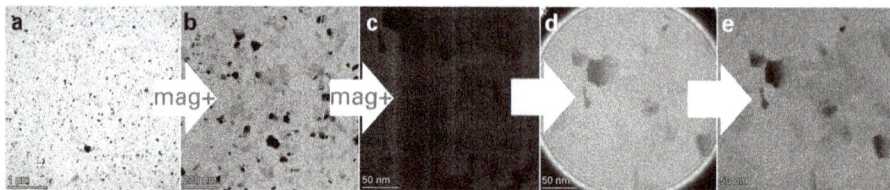

Figure 6.6: Operating the beam during zooming: (a)–(c) During setting higher magnification image became darker, (d) and you need to focus the beam using Intensity and center if using Beam Shift knobs, and (e) and after that expand the beam to cover whole screen or sensor area.

beam – this is due to the imperfections of the optical system, which we compensate by using the central part of the beam.

Efficient operation of the condenser system (continuous focusing and expanding the beam with the Intensity knob along with Beam Shift centering) while changing magnifications is the bread and butter of TEM operation. Focus on doing them quite precise because it helps to obtain good and repeatable results. Speed and freedom in work appear quite quickly, and after some time the researcher operates the beam unconsciously. Importantly, some microscopes have been equipped with automatic beam focusing and centering systems since the 1980s, but they usually have quite limited capabilities (they do not work in the full range of magnifications and/or beam intensity), so you must first polish your own skills and then consider using work support systems (which, however, often work much worse than an efficient operator). For this reason, operation of the condenser system will not be left off the list of necessary skills for a long time (especially if you want to precisely control the experimental and imaging conditions).

6.10 Inserting and centering the objective aperture

There are three main mechanisms of image formation: scattering contrast, diffraction contrast, and phase contrast. With each of them, we are able to affect our imaging using the objective aperture. The smaller the objective aperture inserted into the system, the more diffracted and/or scattered beams we are able to remove from our field of interest. It is necessary to know that diffraction and scattering phenomena occur constantly in our microscope, and the diffraction image describing them is constantly present at the focal point of the objective lens. This is (usually) the location of our objective aperture and its presence in this place gives us unique imaging opportunities, and at the same time makes it difficult to build a microscope because it must be only a few millimeters or even less from the holder and pole piece.

All the time different image formation mechanisms coexist in our TEM. We can control them, [i] among others, by choosing the size and position of the objective aperture

Scattering contrast – it is the mechanism of scattering electrons near atomic nuclei. It dominates in imaging biological and amorphous samples and assumes that for the same preparation density, electron scattering increases with thickness. For the same sample thickness, the shade of gray will become darker when the mean atomic number is higher.

Diffraction contrast – this is the mechanism related to Bragg's law, which describes the deflection of an electron wave on periodic structures in a material, usually the planes of the crystal lattice. It depends less on thickness and atomic number and is strongly correlated with the orientation of the sample. For this reason, when tilting a polycrystalline specimen, we see that individual crystallites become alternately brighter and darker as a function of the angle.

Phase contrast – among other effects, this is the mechanism supporting the imaging of atomic planes or even atomic columns in HRTEM via lattice fringes. It allows us to transmit information about characteristic frequencies through our optical system and show interference between them.

When we insert an objective aperture into our microscope (it is worth starting with the largest available one), in the best-case scenario it will be more or less in the right place. If we are unlucky (or the previous operator simply used another hole or moved the aperture into some specific position) the image will become completely dark or with only a few bright areas. In such a case, we can try to initially find the aperture hole by moving it (using manual knobs or motors), but we should perform this calibration in such a way that it is easy to return to the starting point if we are not successful. The second method is to start the diffraction mode, which displays on our screen an enlarged image of the focal point of the objective lens (i.e., the plane in which the diffraction pattern and at the same time our objective diaphragm appear). If the aperture is moved out, in this mode we will observe a bright spot of the central, nondiffracted beam (it should be located in the center of the image, and each microscope is equipped with the function of centering it), and around it – diffraction points or rings and a blurred halo of scattered electrons. After re-inserting the aperture, we will notice that it is visible in the diffraction image and for standard BF observations it should be located centrally and symmetrically. If you don't see any trace of the aperture hole, reduce the camera length parameter, which will widen the field of view. If the hole is located far from the beam axis, it may be very difficult to see (because only a small part of the beam is deflected so strongly that produce some background signal). In such a case, I recommend working in the dark room with a fluorescent screen and patiently looking for the hole. Sometimes it happens that an unwary operator moved the mechanism very much so that, for example, aperture No. 1 is in the place typical of aperture No. 2, and aperture No. 2 is in the place of aperture No. 3. This type of situation can be recognized by the unusual absence of first or last apertures hole. Of course, after spotting the aperture, you should move it to the center of our diffraction pattern.

When working with single crystalline but dispersed nanomaterials, you may encounter a situation in which dark nanoparticles are accompanied by bright "cousins," which, in the in-focus position, overlap with the dark companion and, in the defocus position, move away from it, each in a slightly different direction. This is caused by the presence of dark field (DF) images superimposed on the image of the undiffracted (central) beam. In such a case, we can get rid of "cousins" (or "ghosts") by inserting and centering a suitably small objective diaphragm that will cover only the central diffraction spot. An example of images with and without "cousins" is presented in Figure 6.7.

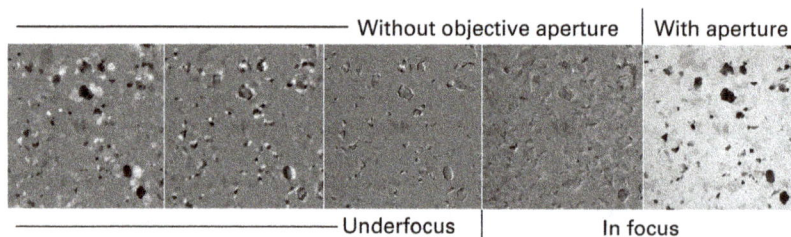

Figure 6.7: Comparison of polycrystalline sample images obtained without and with objective aperture.

Importantly, a contaminated and/or improperly centered objective aperture may introduce aberrations into our microscope that decrease its imaging quality. A quite typical situation is also that after introducing a different objective aperture, the astigmatism of the image needs to be corrected again (which we will talk about in a moment).

6.11 Controlling the focus position and image astigmatism

Another everyday part of working with TEM is focusing and correcting image astigmatism. Both concepts are largely related to each other – a very poorly set astigmatism makes it difficult to discuss the direction and defocus size, and the correct setting of astigmatism requires working with right defocus values.

In the example below, we will use the image of a perforated, amorphous carbon film. This is an example that will be easy to adapt for observing nanomaterials, but at the end of this section, we will also discuss proposed methods for working with other types of samples.

Assuming that in Section 6.8 we set the height of the sample quite precisely using the Wobbler mode, our image should be close to the in-focus position. Shifting the focal plane position upward (usually associated with a clockwise movement of the Focus knob) puts the image into a state of increasing overfocus, and moving the plane down (counterclockwise rotation) puts the image into increasing underfocus. As a rule, a small underfocus (and increasing as the magnification decreases) is the optimal position for most simple observations. It causes a delicate, bright border to appear around dark image structures (e.g., nanoparticles against a carbon film background), improving contrast (Figure 6.8). Its effect can be compared to the "unsharp mask" filter in image correction software. In turn, the overfocus-position will cause a dark rim to appear around the same nanoparticle, accompanied by the brightening of the nano-structure itself. This type of image is widely considered illegible and misleading (Figure 6.8). A hole in an amorphous carrier film will behave similarly – in the underfocus position, a bright, contrast-increasing stripe will appear on its inner edge, and in the overfocus position, a dark stripe will appear on its inner edge (Figure 6.8). Therefore, we can consider working in a small underfocus as the rule of thumb. For this reason, a simple method to obtain the appropriate focal plane position will be to first move into in-focus position (with the support of the Wobbler mode, visually or with the help of the fast Fourier transform (FFT) discussed in a moment), and then move to a slight underfocus.

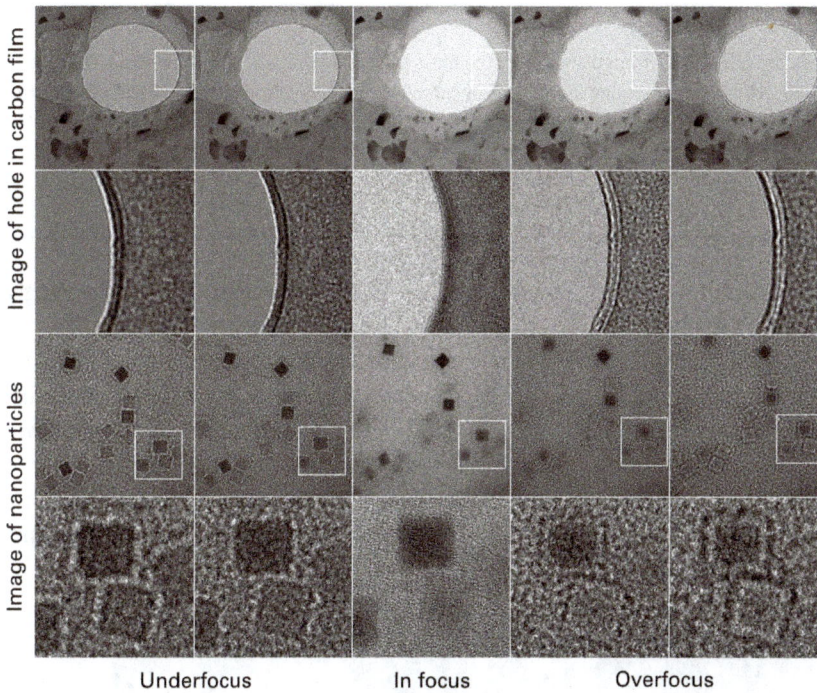

Figure 6.8: Comparison of under-, in-, and overfocus images of hole in carbon film and nanoparticles. The right focus position for most applications is slight underfocus, with delicate bright border around inner part of hole or outer part of nanoparticles. Focus step between images was 1 µm, which is relatively high value, but clearly show differences in defocus.

If we want to image lattice fringes from atomic planes, we can use the Scherzer focus, expressed by the formula:

$$d_S = -\sqrt{\frac{4}{3}C_s\lambda,}$$ (2)

where: d_s – Scherzer focus value (negative, so underfocus), C_s – spherical aberration of the objective lens, λ – electron beam wavelength (nm). For an acceleration voltage of 200 kV and C_s of 1 mm, the Scherzer focus will be approximately 58 nm. When imaging using this defocus value, to put it simply, we maximize phase contrast, facilitating high-resolution imaging of lattice fringes from atomic planes and/or atomic columns (which we usually equate with atomic resolution). Due to the fact that Scherzer defocus is not easy to quickly calculate, some systems have a support system for setting this specific value.

---

i A word about "atomic imaging" in TEM

If in TEM you observe fringes from planes or, in the most favorable situation, images of points from intersecting planes, we are not observing atoms directly but only the effect of the interference of the electron beam with atoms of the crystal lattice, which creates a characteristic image, directly geometrically related to the parameters of the crystal lattice. This does not mean that the planes and atoms are in the exact places you are observing; they may be out of phase with the image. But still, the ability to observe an atom-like image is incredibly fantastic and impressive!

An extremely useful feature of modern, sensitive CCD/CMOS cameras (and increasingly also cameras observing a hidden fluorescent screen) is the ability to perform FFT in real time. This allows us to observe Thon rings in defocus positions on amorphous samples (or on samples laying on a supporting, amorphous film), but without a simple distinction between under- and overfocus positions. Aiming for the in-focus position, we expand Thon's rings so that in the central position we see no ring (Figure 6.9). If we have doubts about which direction leads to underfocus, just observe the microscope's defocus value while turning the Focus knob. The underfocus position tends toward lower and/or negative values.

| Underfocus | In focus | Overfocus |

Figure 6.9: FFT images of amorphous carbon supporting film in different defocus levels. As the focus approaches the in-focus position, the Thon rings become wider and disappear. Focus step was approximately 1 μm.

Once we have the basics of our focus position, we need to return to the concept of astigmatism. As a simple definition, astigmatism causes parts of the beams focused in two perpendicular planes to be focused at different points (so focal points are different at different beam orientation). Therefore, it may turn out that if we image a hole in the supporting film, its two fragments show the features of a light underfocus stripe and the places at another angle – the features of a dark overfocus stripe (Figure 6.10). For the completely defocus position, uneven thickness of the fringes can be observed. In the older types of microscopes, the standard was to set the astigmatism by homogenizing the thickness of the bright fringe in a slight underfocus (Figure 6.10). Unfortunately, the matter became more complicated when we significantly changed the astigmatism of the system by installing an objective aperture or a sample with noticeable magnetic features. In such a case, I recommend using the approach that I already proposed when correcting condenser astigmatism. This time we will alternate between the Focus, Obj Stig X, and Obj Stig Y knobs, each time maximizing the visual sharpness of the image. This is not an easy task, so I suggest turning the knob hard each time so that the image looks "obviously bad." Then we pass the best position and keep spinning to find the similar "obviously bad" position again. The local optimum is exactly between these two values, even if it is not an absolute optimum. In this way, by correcting the two axes of the stigmator and the focus position multiple times in a row, we can achieve a satisfactory setting of focus and astigmatism on most simple samples. All that remains is to introduce a minimum defocus and take the exposure.

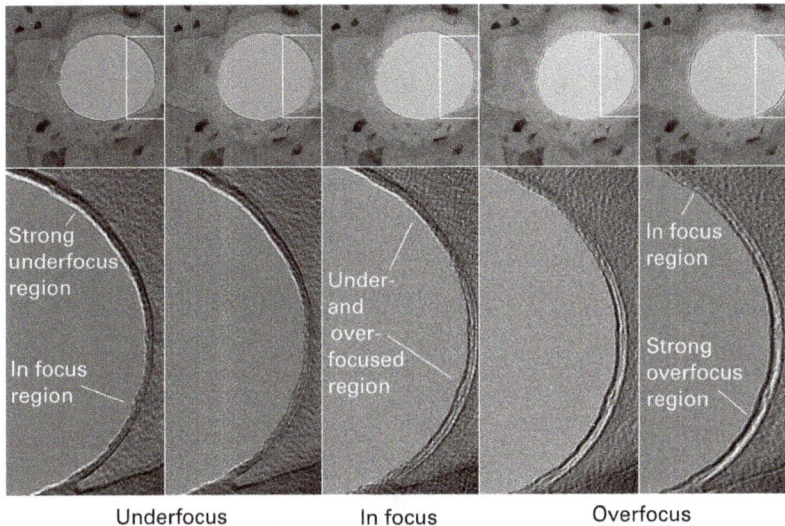

Underfocus	In focus	Overfocus

Figure 6.10: Uneven image of light and dark fringes associated with under- and overfocus when the image is astigmatic and needs correction.

The matter is greatly simplified by the use of Thon rings on an amorphous sample or film. In this case, the image is free of astigmatism when Thon's rings are symmetrical and round. If they are ellipses, simply activate the astigmatism correction on the multifuncional knobs or use the dedicated knobs. It is worth noting that astigmatism can change slightly even as a function of the position of the Focus knob. For this reason, using CCD/CMOS cameras, we are able to minimally correct astigmatism just before each exposure.

It is worth remembering that even when using the benefits of FFT, we may find ourselves in a situation where setting astigmatism will be difficult. This happens when, instead of Thon rings, we see an X-shaped FFT image. This means that for the current defocus value, astigmatism is exceptionally significant. My advice in this situation is to increase the defocus value until the FFT recovers Thon elliptical rings (Figure 6.11). Then we initially correct the astigmatism, reduce the defocus, and correct the astigmatism again until we reach the in-focus position. What is easy to notice is that in focus it is difficult to correct astigmatism using FFT because the Thon rings are invisible. This is completely normal. For your convenience, in Figure 6.11, you can see example "maps" of Thon rings as a function of defocus and astigmatism in one axis.

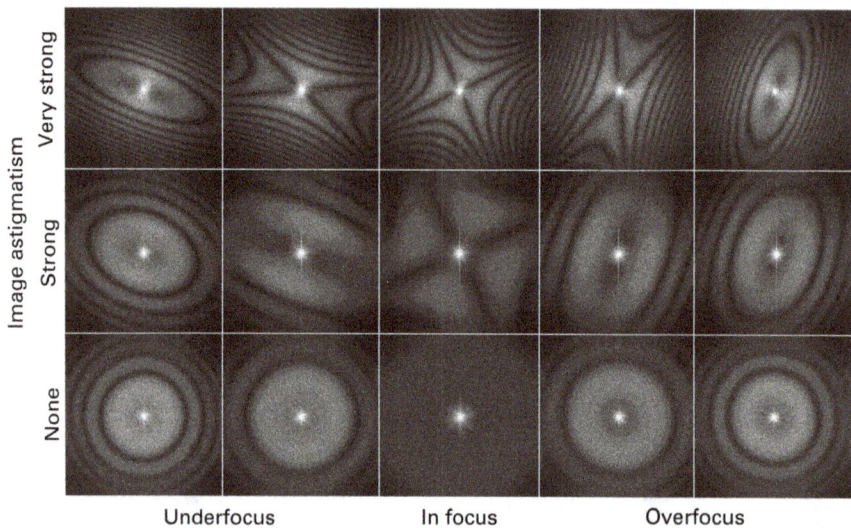

Figure 6.11: FFT image for different defocus levels and degrees of astigmatism. You can notice that if you see cross patterns, they will be easier to correct in higher defocus value.

6.12 Taking TEM images using digital cameras

Much of the preparation for taking micrographs has been described in the previous section. Importantly, not every frame has to be prepared perfectly. If you have made a number of observations at high magnifications after setting the right stigmator values and want to make several exposures at medium and low magnifications, you probably do not need to correct the astigmatism further (and even some slight imperfection will not be visible in the image). The parameter we must take into account is the exposure time of the image. If it is too short in relation to the sample illumination (Figure 6.12), the image will be dark or noisy (with a small signal-to-noise ratio). If it is too long, it may turn out that we will preserve the image of sample drift (analogous to a blurred photo taken with a camera). Drift is especially visible shortly after changing the specimen (when the sample holder homogenizes the temperature with the rest of the microscope) and when working on flexible supporting substrates (such as low-mesh-value carbon film, especially near film cracks). If possible, an easy remedial action is to observe the sample near the bar of the supporting mesh. It may be a rule of thumb to assume that if sample drift is visible to the naked eye on the camera preview screen, the micrograph taken during a 1-s exposure will probably be blurred. If you work with a CCD/CMOS camera, try to spend some time fine-tuning the parameters of exposure time and beam intensity within the frame that will satisfy you. Remember that changing the condenser parameters (Intensity knob and possibly the Spot Size parameter, which regulates the defocus of the first condenser lens C1/K1) may affect sharpness and astigmatism. In turn, long setting of exposure, beam, and focus parameters in one place may cause accelerated contamination (Figure 4.3b). For this reason, if the area of interest becomes contaminated, it is worth changing the observation site a moment before taking the image.

You should also avoid a situation in which the CCD/CMOS sensor becomes oversaturated due to excessive high beam intensity and/or long exposure time. This is easily observed by the reduced tonal range of the image, the lack of information in the brightest areas, and the visible fiber honeycomb structure (Figure 6.12).

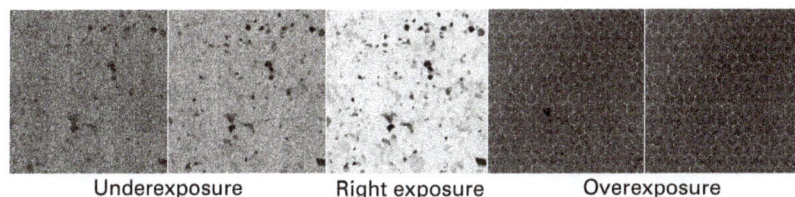

Underexposure   Right exposure   Overexposure

Figure 6.12: Effects of underexposure under- and overexposure on CCD/CMOS cameras.

6.13 Getting ready for sample change

If you plan to replace the sample, an important rule is to leave the microscope in a condition that will make it easier for you to work with after the replacement. Most producers require and recommend that before removing the sample, ensure that all

X, *Y*, and *Z*-positions and rotation angles α and β (for holders that provide double tilt functionality) are reduced to zero. This prevents unnecessary collisions with the elements inside the microscope and goniometer.

Before zeroing the goniometer position, it is worth ensuring that it will be easy to start observations after replacing the sample. For this purpose, it is good to set the lowest practical magnification, and for magnetic samples we use the Lorentz microscopy mode or low magnification (LM) mode to minimize the current in the objective lens, and thus the magnetic field and the possibility of throwing the sample out from the holder. Then we expand the beam using the Intensity knob and center it to cover the entire field of view. Thanks to this, we will not have to wonder what is obscuring the image in a situation where, after replacing the sample, we see only darkness (quite similar to procedure described in Section 6.2).

It is also necessary to ensure safe operating parameters of the electron gun. Sample holder removal is the second moment (after installation of the holder) when the vacuum is most often disturbed. Therefore, we want to ensure that any air leakage into the microscope does not negatively affect our electron gun. Many designs recommend closing the gun valve or turning off thermal emission if the microscope does not have such a valve. We have already described the issue of emission handling in Section 6.2 of our procedure. If our microscope is equipped with a retractable EDS detector or other delicate accessory, it should be moved to a safe position in accordance with the instructions or manufacturer's recommendations. It is worth doing the same with the objective aperture, which is very close to the holder.

6.14 Sample change

After confirming once again that the microscope is in a condition suitable for pulling out the specimen holder (stage in zero position, valves closed, electron gun safe, detectors, and aperture extended), we can pull our holder out of the column in the opposite direction from the one described earlier. The exact methodology varies from manufacturer to manufacturer, so we won't describe the process in detail. However, I encourage you to refresh your knowledge on the content of Sections 6.5–6.7.

6.15 Turning off the microscope

First of all, it is worth remembering that not every TEM needs to be turned off after work. Devices with emissions more demanding than thermal emissions from a tungsten cathode usually work well when operating 24/7. Typically, the procedure for full and safe shutdown is described in the user manual. However, before starting it, it is worth following the rules described in Section 6.13 to make it easier to start work the next day.

Devices that use an ACD usually require a bit of special treatment at the end of the work day, even if it doesn't involve a full shutdown. Usually, in the manual we can find a procedure for heating the anticontamination trap, called ACD heating, ACD heat, cryo cycle, or similar. It involves the user removing the remaining liquid nitrogen from the dewar (using a pump, an additional heater, or removing and emptying the container), and then the cold trap inside the microscope is heated using an additional heater or using the temperature surrounding the microscope. If the device uses ion pumps, they are often turned off at this point, and diffusion or turbomolecular pumps remove the gas molecules and impurities adsorbed on the trap. This type of action extends the life of ion pumps and improves vacuum conditions because the mentioned pumps are not good at removing water vapor from the inside of the column.

When finishing work, it is worth considering performing other occasional maintenance activities, such as draining the water from the bottom of the compressor tank, that compresses the air to operate the valves (if your microscope uses pneumatics). This operation is usually performed weekly or monthly to remove condensed water from the bottom of the tank. This slows down its corrosion and improves the operating conditions of the valves. The author once worked on a device in which such an operation had not been performed for over 10 years, and the vacuum valves operated using water pressure rather than air pressure, and big part of the valves was not moving due to the corrosion process.

If your microscope is equipped with water-cooled diffusion pumps or turbomolecular pumps, ensure that the cooling water circulates for at least 15–20 min after the microscope is turned off. This is the time required for the pumps to cool down to the appropriate temperatures and not locally boil the cooling water. It is worth using the time waiting for the pumps to cool down to tidy up the workplace, recording, saving, and copying the images or to study the literature or the device's operating instructions.

Take-home messages

– Before starting work, make sure your microscope is in good condition. Check the vacuum, emission parameters, and other factors indicated by the user manual.
– It is a good idea to check the shape and position of the beam before installing the first sample. This allows for easier troubleshooting if we get a dark image.
– In the case of ferromagnetic samples, it is sometimes recommended to install and remove them from the column in a specific imaging mode such as low magnification (LM) or Lorentz microscopy mode.
– After installing the sample and starting observations, first, it is worth setting the cheese defocus, adjusting the appropriate sample position on the Z-axis, and verifying the correctness of the rotation center.
– When removing or introducing the specimen holder, make sure that your electron gun is safe (gun valve closed or thermal gun emission turned off), the goniometer is in the basic position, and other conditions described in the user manual are met.
– Before the preparation or ending work, make sure that the device is left in the condition in which you would like to find it.
– If you have problems with some procedures (focusing, apertures, astigmatism), remember that you can refer this chapter again.

Literature

[1] Fujita H. History of Electron Microscopes. Int. Congr. Electron Microsc. 1986.

[2] Hawkes PW. Advances in imaging and electron physics: Growth of electron microscopy [Volume 96]. Academic Press; 1996.

[3] Yoshimura N. Historical evolution toward achieving ultrahigh vacuum in JEOL electron microscopes. Tokyo: Springer Japan; 2014.

[4] Williams DB, Carter CB. Transmission electron microscopy: A textbook for materials science. 2nd ed. New York: Springer; 2008.

[5] Reimer L, Kohl H. Transmission electron microscopy: Physics of image formation. 5th ed. New York, NY: Springer; 2008. (Springer series in optical sciences; vol 36).

7 Using a microscope – research using diffraction

In the previous chapter, we discussed the basics of practical transmission electron microscopy imaging. In this section, we will discuss the issue of making selected area electron diffraction (SAED) patterns and bright and dark field observations. This is a technique that has been known for over 60 years, but it still offers unmatchable possibilities in terms of detecting the presence, distribution, size, and crystallographic orientation of individual crystals of our sample [1]. To effectively use the knowledge described below, it is worth having basic knowledge of crystallography [2]. However, if you miss it, nothing is lost. After your first attempts at working with diffraction on TEM, you will see what knowledge you lack the most.

When starting to work with diffraction techniques, you should referb the experiment shown in Section 6.11.

7.1 Selecting a good location for electron diffraction pattern observation

In Section 6.10, we discussed the basics of using and centering the objective aperture of our microscope. If you are working with a crystalline sample and have an objective aperture inserted (the smaller it is, the more visible it will be), then for areas of similar thickness, there is a greater chance of observing "good" diffraction in places that are darker (so the diffracted beams are stopped by objective aperture). Sometimes these types of areas form stripes related to the local bending of the sample (called bending contours). If you encounter a spot where several of these lines intersect in a spider-like pattern, or you observe a wide, dark region (that is not darker just because it is thicker), this is a likely location for observing promising electron diffraction (Figure 7.1).

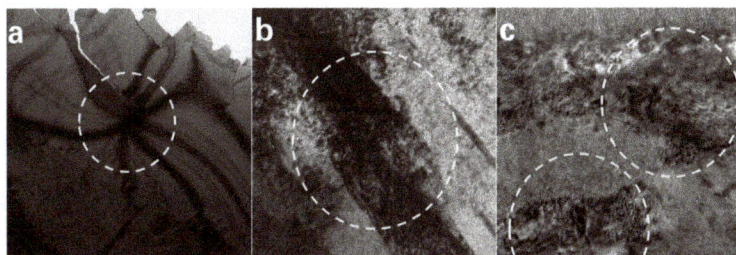

Figure 7.1: Examples of places where satisfactory SAED patterns can be expected: (a) the place where the bending contours intersect in the spider-like feature, image width 10 μm; (b) a clearly darker grain, image width 1 μm; and (c) crystallites slightly darker than the surrounding ones despite having the same thickness with image width 500 nm.

https://doi.org/10.1515/9783111317014-007

A different situation occurs when we have many small crystallites in our field of view and/or we do not care about examining individual crystals, but rather the broader picture of the presence of individual phases. In such a case, we will try to make a diffractogram from many crystals and the sum of the individual diffraction patterns (Figure 7.2b) of which will create a ring diffraction pattern (Figure 7.2a).

---

i | **There is no such thing as good or bad diffraction pattern**

Our TEM is a diffraction machine and does not evaluate our choice of observation site. However, it is believed that a "good" single crystal diffraction consists of at least eight symmetric and regular diffraction reflections and a central spot (Figure 7.2b) of the phases we expect in our sample.

If our sample does not show crystalline features, we can expect to observe a broad "halo" in the diffractogram (Figure 7.2c). Importantly, a similar situation will occur when we test a material that is extremely sensitive to electrons, it may undergo amorphization even before you note the first observations.

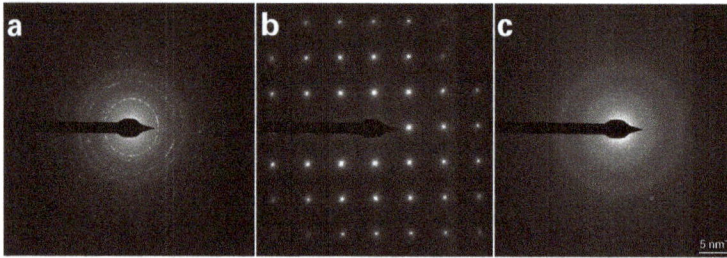

Figure 7.2: Typical SAED pattern images: (a) narrow rings from many randomly oriented crystals within the diffraction aperture; (b) single crystal diffraction with regular reciprocal lattice points; and (c) wide halo-blurred rings typical of an amorphous site.

7.2 Setting parameters for proper diffraction pattern observation

Electron diffraction is based on the same physical principles as, for example, X-ray diffraction (XRD). The phenomenon of diffraction of X-rays or electron waves on the planes of the crystal lattice is described by the so-called Bragg's law:

$$n\lambda = 2d \sin \theta \qquad (3)$$

where n is the order of deflection (usually one or a small natural number), λ the wavelength, d the interplanar distance, and θ the angle of incidence of the beam relative to the plane.

This means that for a specific wavelength, angle, and interplanar distances, a situation can be observed in which the conditions of Bragg's law are met and diffraction occurs [3]. Fortunately, for many decades TEMs have been designed in such a way to enable the observation of diffraction phenomenon. Interestingly, if you take a ring diffractogram and perform a circular plot (distribution of average intensities along the radius, relative to the central spot), you will obtain a spectrum very similar to the XRD spectrum. However, it will have much wider peaks, and this is caused by a number of imperfections of TEM as a diffraction instrument. For this reason, while for XRD the error in measuring the interplanar distance is a small fraction of a single per cent, in TEM, the error can reach 5–10% and often depends on the operator's work. However, to ensure maximum precision in working with diffraction on TEM, several important assumptions must be considered:

- Calibration of the camera constant (or more simply – scale bar on diffraction) should take place in the same, unchanging, and constant parameters as the actual diffraction examination;
- Camera constant calibration should be regular. As months and years pass, this value may change, especially on older microscopes;
- Camera constant changes significantly in the image and diffraction defocus function (Figure 7.3). For this reason, both calibration and testing should be performed at a constant defocus value, preferably equal to zero (zero focus position, see Section 6.8). Of course, we can't always set the perfect zero focus position using only the Z axis, but you should still minimize defocus;
- The sample should be at or close to the in-focus position;
- The diffraction aperture should be centered;
- The condenser beam should be parallel (not convergent or divergent). We will discuss the exact beam positioning in Section 7.5;
- The microscope should operate in an imaging mode supporting diffraction studies (often called SA, BF, or unmarked – check the user manual).

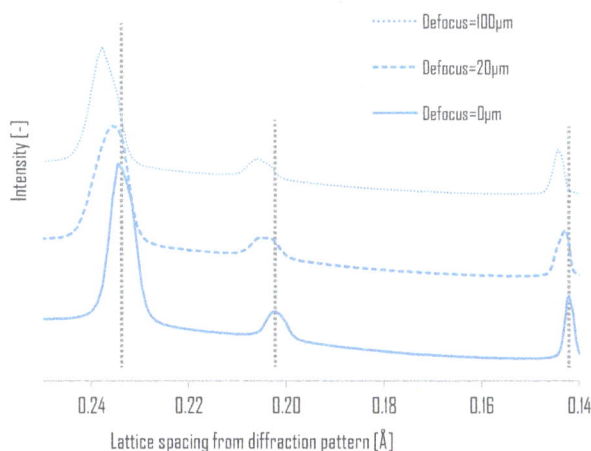

Figure 7.3: Diffraction peaks shift at different defocus position, shown on circular plot of polycrystalline SAED. This is one of the motivations for calibration and further work at zero defocus position.

7.3 Inserting and centering the diffraction aperture

When we want to observe a diffraction pattern, we should place the area of interest in the center of the frame using the XY sample shift axes, select a magnification that allows viewing a slightly wider frame, center the condenser beam, set the in-focus position, and then insert the appropriate size of the diffractive aperture. It should include the area of interest and not overlap too much with the surrounding crystals (for studies of single crystallites) or cover dozens and hundreds of individual nanoparticles or crystallites (for polycrystalline diffraction). For a better selection of the place to perform SAED the microscope is equipped with multiple diffraction apertures of different diameters. The diffractive aperture should be carefully centered relative to the screen or camera. After installing it, you may have difficulty judging whether the area of interest is still inside, which is why we took care of this before. If you want to be sure, remove the aperture for a moment and check the position of the sample. If you have a digital camera, you can additionally take microphotographs without and with an aperture during the first experiments, which will help further interpretation of the research. For routine work, it is usually enough to mark the place where the diffraction was performed in one of the wide area micrographs or in the notes.

7.4 Starting diffraction mode and removing the objective aperture

Due to the high intensity of diffraction peaks, work with diffraction should be performed on a fluorescent screen and use a CCD/CMOS camera only for image acquisition. Starting diffraction pattern observation you should enter diffraction mode (DIFF or similar button/switch) and simultaneously remove the objective aperture (if you still had it inside). If your microscope is initially centered correctly, you should see a diffraction pattern of your sample, perhaps still a bit blurry (blurry diffraction pattern means, unlike a regular image, that it could consist of sharp points but in the wrong places). You can choose the appropriate camera length (values in the range of 600–900 mm could be good starting point) and center the diffraction image in relation to the screen. Usually, dedicated knobs or appropriate multifunctional knobs are used for this purpose (and they operate some beam deflectors available inside the column). In older devices (1970s and older), diffraction centering was sometimes achieved by physically moving one of the projection lenses by dedicated, mechanical knobs.

---

i Diffraction camera length

When it comes to choosing the diffraction camera length, it describes the size of the diffraction image. The longer the camera length, the larger the diffractogram on our screen or camera will be. This makes it easier to recognize reflections close to the central point at longer camera lengths and to observe distant diffraction rings at short camera lengths. However, it is worth remembering that the ratio of the size of our objective apertures to the diffraction size will remain unchanged. Moreover, for large camera lengths it may turn out that our central point is significantly larger than the beam stopper.

7.5 A few words about sharpening – image, diffraction, and aperture

Diffraction sharpening is a broader concept than it may seem at first glance. Basically, almost any diffraction image can be brought to the correct focus using the Diffraction Focus knob (DIFF Focus or similar), but it is worth remembering that the size of our diffraction will also depend on the position of the imaging defocus (observed in the image, other than the diffraction focus). For this reason, in the previous point we made sure that our sample was in the in-focus position, as close to the zero focus position as possible. It can be assumed that the appropriate DIFF Focus position is when the central diffraction reflex is the smallest and sharpest. You may find that you need to use a binocular for this setting. Importantly, the position of the DIFF Focus will change as a function of the convergence of the condenser beam. If you turn the Intensity knob you will notice that the sharpness of the diffraction changes. In the "quick and easy" approach, it is usually enough to ensure that the intensity is sufficient to perform the diffraction exposure at a predetermined (and fixed) time. This usually requires a fairly widened condenser beam, for such a setting, little changes to the Intensity knob slightly affect the sharpness of the diffraction. Sometimes optimum diffraction focus condition is also available as factory-adjusted neutral diffraction focus (with reset or zero focus button).

However, if we want to perform a diffraction study in the state of the art, we should make sure that our condenser beam is as parallel as possible. For this purpose, when observing the diffraction image, we should introduce one of the larger objective apertures into the column so that we can see its edge in the diffraction image. Then we need to set the Intensity knob so that the edge of the objective aperture is as sharp as possible. This is our parallel beam position. Now, without modifying the position of the Intensity knob, we sharpen the central point of diffraction using DIFF Focus. If the brightness of the diffraction image is too high, we can change the spot size (and repeat the entire procedure from centering the beam and sharpening the image). If it's too small, you probably just need to increase the exposure time on the camera. However, if our microscope allows it, it is worth using a beam blanker, a small cover of the central point. It allows you to protect the detector against excessive exposure. Even if we do not damage the scintillation material, it may show a memory effect, which will make subsequent micrographs look worse.

7.6 Diffraction, astigmatism, and its correction

The diffraction image can also show astigmatic features and has its own astigmatism correction systems. Very high diffraction astigmatism may result in polycrystalline elliptical diffraction rings. In such situations, if diffraction astigmatism could not be corrected using stigmator, I recommend contacting the service and having the column thoroughly cleaned.

Correcting astigmatism is easiest to do in a similar way to diffraction pattern sharpening. We will try to minimize the central point of diffraction. If you have a problem with this, I recommend starting from noncorrected position (with wide spot), turning the knob a lot so that the central point looks a bit better and then once again wide, and after that, finding the optimal position again. If you are examining a polycrystalline specimen (or, e.g., a diffraction calibration specimen), you will also notice that as the diffraction astigmatism improves, the rings become narrower and sharper (Figure 7.4).

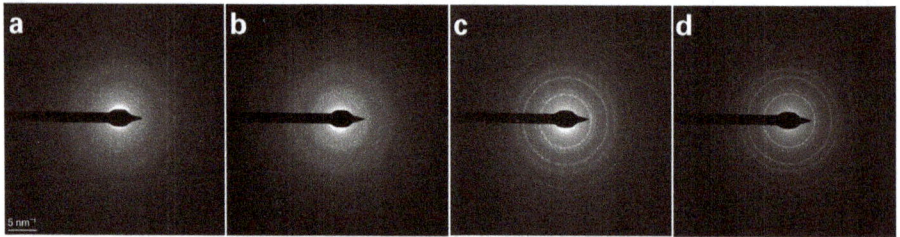

Figure 7.4: Typical images visible during focusing polycrystalline diffraction: (a)–(c) defocussed SAED pattern and (d) sharp SAED pattern from polycrystalline Al sample.

7.7 Bright and dark field observation – introduction

If we take a sharpened diffraction pattern and start defocusing it with the DIFF Focus command, we will see that the diffraction spots get larger and larger until at some point each of them becomes a separate image (Figure 7.5). If we now move the sample, we will see that its image moves in each of the "mini-images." This is simple proof that each diffraction reflection contains full information about the places that caused diffraction in a specific way (created a given diffraction point). If we have points corresponding to different phases of our sample, we can easily indicate the specific location, shape, and arrangement of crystals of this phase in a given orientation.

In Section 6.10, we discussed the basic insertion and centering of the objective aperture. This time our actions will be very similar, but they will aim to alternate operation in the bright and dark field modes. To do this, we need to make sure that our microscope is preset in a way that will allow us to make these observations. We will discuss performing dark field observations using two methods – the aperture shifting

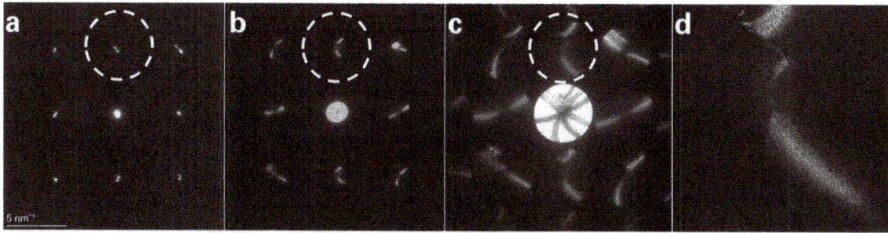

Figure 7.5: The proof that each SAED point contains information about the DF image: (a) focused, single crystal SAED pattern; (b) slightly defocused SAED pattern; (c) strongly defocused SAED pattern, showing ordered BF/DF images limited by a diffraction diaphragm; and (d) DF image, taken with objective aperture introduced in place marked with circles.

method and the beam tilting method. The first of these methods could sound easier, but practically has many flaws. It offers lower resolution, problems with focusing, and requires repeated manual shifting of the objective aperture. The beam tilting method requires a more precise preset of the microscope, but in return it offers great convenience, easy focusing, and higher resolution [4]. So the choice seems simple.

7.8 Obtaining a dark field image using the moving aperture method

In this method, we insert an objective aperture into the diffraction image. If it includes the central spot, we will be dealing with a bright field image (possibly with "ghosts" of dark field images, if some diffracted spots are also inside the aperture). If we do not have a central reflection in the aperture, but only diffraction reflections, we will be dealing with a dark field image. The choice between both modes is physically accomplished by moving the aperture. It is worth choosing the aperture size that includes only one of the reflections we are interested in. If you want to examine intermetallic precipitates in metal alloys, you will usually have to choose the smallest apertures (Figure 7.6), especially if you choose reflections close to the central point (unfortunately, the closer to the central point, the more "valuable" the images from diffraction points tend to be). Returning to imaging mode, you will see the appropriate bright or dark field image. You'll probably have to pull out the selected area (diffraction) aperture in the meantime, which covers big part of the field of view. It is worth remembering to put it back in place before taking another look at the diffraction pattern.

In dark field images, the bright spots are those that created a given diffraction reflex. Don't be surprised when you move the Intensity knob or move the condenser beam by Beam Shift, some areas change their contrast – any change in the electron beam parameters that affects diffraction (as described in the previous points) may affect the dark field image as well.

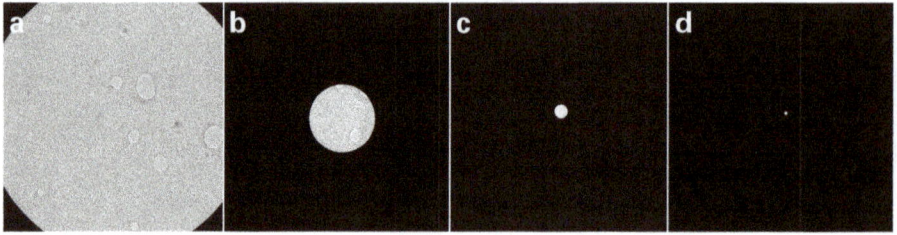

Figure 7.6: Exemplary set of four different diffraction apertures, available in specific TEM, imaged on the same magnification. Diameter of (a) 800 μm, (b) 200 μm, (c) 40 μm, and (d) 10 μm. Each frame width is 15 μm.

You can also look at the diffraction image with the objective aperture retracted and see in which directions and how far you need to turn the Objective Aperture Shift to observe a specific reflex and/or central point (Figure 7.7). Then, in the image preview, you can repeat this aperture movement to switch between the bright and dark field image. Don't be surprised that at some point you will go "beyond the edge" of the beam – the image of the beam that we know and center for most of the work is a bright field image and we are naturally more used to seeing it. It is also completely normal for dark field images obtained using the moving aperture method to shift sideways when focusing with the Focus knob, as if the rotation center has not been set properly. This is a manifestation of the geometric imperfections of the moving aperture method.

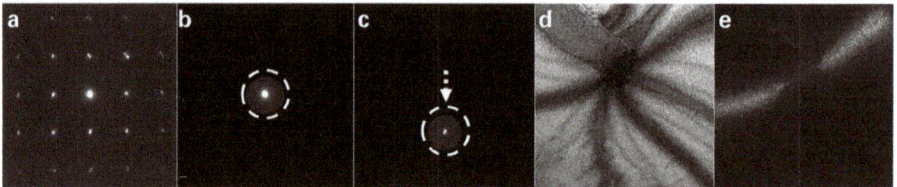

Figure 7.7: Procedure for DF imaging using moving aperture method: (a) centered SAED pattern, (b) place of aperture for BF imaging, (c) aperture shift to obtain one of the DF images; (d) BF image with aperture in position marked on (b), and (e) DF image with aperture in position marked on (c). BF/DF images width is 4 μm.

7.9 Obtaining a dark-field image by tilting the beam

This type of method is characterized by a much higher quality of the obtained images and the ease of switching between a bright and dark field image, without physically moving and centering the aperture. In this case, the aperture remains centered – perfectly in the middle – and we move the diffraction pattern in its plane by tilting the condenser beam. To do this, make sure that you are close to the zero focus and in-focus positions and check the correctness of the "pivot points" parameter (also called Tilt Align, DF Align or similar). This setting ensures that when tilting the condenser beam, it does not visibly move away from our area of interest. Usually, to correct it, you need to be in the zero focus position. After correcting the rotation center (described in Section 6.8 and in Figure 6.5), focus the condenser as much as possible, center it, and then run two pivot point adjustments in the X and Y axes, and use the appropriate knobs to achieve a situation in which the most intense parts of two flashing beams (often looking like a comet nucleus) coincided with each other (Figure 7.8). After making the correction and expanding the condenser, we are ready for our observations.

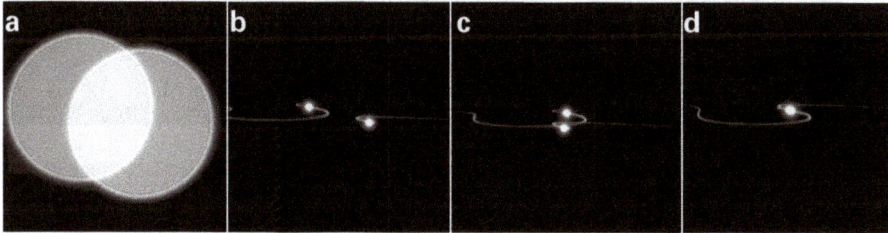

Figure 7.8: Procedure for aligning one of the pivot points. Note that you shouldn't worry about the narrow trace of the beam: (a) too wide beam, (b) focused beam but pivot point is misaligned in both directions, (c) one of the pivot point directions aligned, and (d) pivot point fully aligned.

Your microscope should be equipped with a panel, buttons, or a menu in the software that is used to observe dark field images. First of all, make sure that in both modes, BF and DF, the condenser includes the same area in the image mode, and in the DIFF mode the diffractograms look the same (they are not shifted). If they are shifted, in DF mode, reset the Beam Tilt position (if the microscope allows it) or manually use the DF Beam Tilt command (DF Tilt or similar). This way we bring the diffraction in DF mode to a position analogous to the BF position.

We start the DF observations using the beam tilting method in the diffraction mode. We have the diffractive aperture inserted and centered, the objective aperture removed, and the diffraction pattern sharpened and precisely centered (centering is especially crucial!) as shown in Figure 7.9a. Then, in DF mode (started by a button, panel, or software), we use the DF Tilt command so that the spot we are interested in reaches the center of the fluorescent screen (Figure 7.9b). If the image disappears, we probably set the pivot points incorrectly and it is worth working more on them. We can then switch between DF and BF modes (still in DIFF mode) and make sure that the central spot in BF mode and our new reflex in DF mode are in the same place. If not, we perform the DF Tilt correction again in DF mode. If everything is correct, we insert the objective aperture in DF mode so that it includes only our specific diffraction spot (Figure 7.9c). In BF mode we confirm once again that the central reflex is inside the aperture (Figure 7.9d). We are now ready to return to imaging mode and possibly remove the diffraction aperture (not objective!). In the BF mode we will see a typical bright field image, and in the DF mode we will see places that have undergone diffraction. Some modern microscopes even offer the ability to select multiple individual reflections in DF mode and easily switch between them. I recommend reading more about BF/DF observations in your microscope's manual – they often differ in small details.

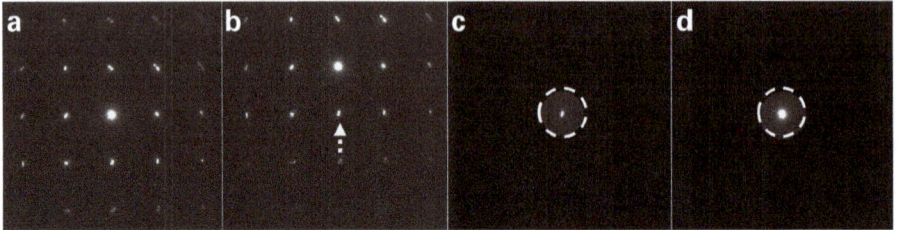

Figure 7.9: Procedure of DF imaging using beam tilt method: (a) centered SAED pattern; (b) tilting the pattern in DF mode, just to center spot of interest; (c) image after introducing objective aperture during DF conditions; and (d) image after introducing objective aperture during BF conditions, please mention higher intensity of BF spot.

7.10 Comparative bright and dark field observations

Effective and efficient performance of BF and DF observations require some skill and experience, but in return it offers a unique opportunity to detect individual phases and crystallographic orientations.

Example imaging using SAED/BF/DF techniques presents a diffraction pattern of two phases (ferrite + cementite) in bainitic steel (Figure 7.10). The (200) reflex of cementite (marked as DF on Figure 7.10a) was used for dark field imaging. The resolved diffraction shown in Figure 7.10a indicates crystallographic relationships: the zone axes of the [001] ferrite and [010] cementite are parallel to each other, as are the (200) ferrite and (204) cementite planes. This indicates a partial coherence of both phases and is a typical example of the use of DF technique (Figure 10c) to detect the presence of a phase that is poorly visible in BF (Figure 10b).

Figure 7.11 shows an interesting DF image from the reflection, which has undergone the so-called double diffraction phenomenon. This means that one of the diffracted beams was diffracted again by another crystal before it left the sample. Figure 7.11a shows BF of the high-nitrogen stainless steel, with plate-like structures seen in the matrix. Figure 7.11b shows DF images of austenite crystal and Figure 7.11c shows DF from plates. However, on the SAED pattern (not shown here due to complicated and noneducational structure) there were also some peaks, which reveal the DF image visible in Figure 7.11d. You can see that the brightest spots are where the two crystals overlap in the observation plane. This means that the reflection occurred only in the place

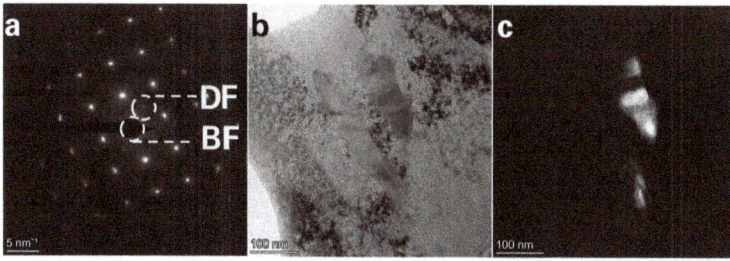

Figure 7.10: Example of SAED/BF/DF data: (a) SAED pattern from ferrite and cementite diffraction in alloy steel, with marked spots for BF and DF imaging; (b) BF image, showing weak contrast from cementite; and (c) DF image from cementite diffraction spot.

where both crystallites were present, so it is the result of double diffraction, from one and the other crystal lattice directly after each other. In the diffraction pattern, such points are the vector sum of both diffraction points typical for both phases (the beginning of the vector – the central point, the end of the vector – the diffraction point). You can also notice brighter structures in other parts of the frame – they are caused by nearby peaks that could not be avoided even with the smallest diffraction aperture. This is also a situation that can happen during everyday work.

In this way, virtual reflections may appear on our SAED that do not match any of the phases. It is easy to see that the bright areas are where two phases occur simultaneously, which confirms the double diffraction hypothesis. Of course, I wish the reader to experience as few surprises and artifacts as possible during research.

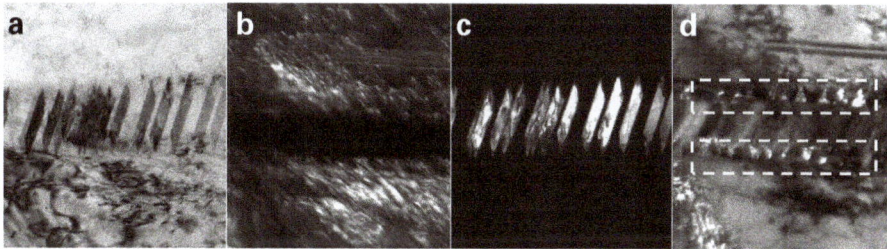

Figure 7.11: Example of BF/DF imaging with double diffraction: (a) BF image of plate-like structures in austenitic matrix in stainless steel, (b) DF image from austenite grains, (c) DF image from plate-like structures, and (d) DF from double diffraction spot, the brightest spots indicate where the two crystals overlap. Each frame cover width is 500 nm.

---

i Take-home messages
- TEM is a diffractive device, and diffraction and scattering phenomena occur continuously in it and can be observed in the first focal plane of the objective lens. It is also an image that we see in the diffraction mode.
- By inserting the objective aperture in the objective first focal plane, we can obtain bright- and dark-field images, and the latter tell us about the presence of places (phases, crystallites) that have diffracted the electron beam in a specific way (created the examined diffraction spot).
- Wide diffraction apertures are helpful for collecting polycrystalline diffractions, and narrow diffraction apertures are helpful for working with single crystals.
- Of the two methods of obtaining a dark field image, the beam tilting method is particularly advantageous, although it requires a slightly more precise aligning of the microscope and the beam.
- Both calibration and diffraction work should be done in the same defocus position, preferably zero defocus.

Literature

[1] Hirsch PB. Electron microscopy of thin crystals. Robert E. Krieger, Malabar, Florida; 1977.
[2] Andrews KW, Dyson DJ, Keown SR. Interpretation of electron diffraction patterns. Springer Science+Business Media, LLC; 1967.
[3] Fultz B, Howe JM. Transmission electron microscopy and diffractometry of materials. 3rd ed. Berlin, New York: Springer; 2008. (Advanced texts in physics 1439-2674).
[4] Williams DB, Carter CB. Transmission electron microscopy: A textbook for materials science. 2nd ed. New York: Springer; 2008.

8 FAQ – common problems and how to deal with them

Here I will discuss some of the most common problems and challenges that operators face during their daily work. The list is not exhaustive, but I hope it will be helpful for your first attempts at simple troubleshooting.

8.1 After inserting the sample, I don't see the image

This is a very typical situation. If the specimen is a nanomaterial on a support grid, it usually helps to move the sample along the X and Y axes. In several instances in the book, I recommend leaving the electron beam in a tested and appropriate state before replacing the sample. After that, the black image could be caused only by the sample shadowing. If you followed Section 6.2 and still cannot obtain an image with the sample, try again to observe the beam without it. If the sample is a thin foil, finding the hole and the transparent area may be problematic (as described in Section 6.8). It is worth noting that in the case of hard magnets, they can slightly deflect the condenser beam from the standard position. In this case, try expanding the beam with intensity knob, lowering the magnification even further, and systematically scanning the sample area with X and Y axes.

8.2 After the magnification change, image disappeared

In such a situation, I recommend immediately returning to the magnification or parameters at which the image was still visible. Focus, center and expand the condenser with intensity and beam shift, because significant beam off-centering may result in difficulty finding the beam at higher magnification (the edge of the beam disappears from the field of view when it is still very dark and poorly visible). If your microscope offers "automatic" condenser positioning functionality, disable this feature. Especially in older designs, it can do more harm than good (and in modern TEMs it still works not as good as experienced operator). Some microscopes remember the position of the condenser at different magnifications in internal memory. You can check if this is the problem by focusing the beam so that you can clearly see its edges and changing the magnification slightly (up or down by one increment). If the condenser spot has moved noticeably, it is due to the imperfection of the microscope's centering. In such a situation, I recommend consulting the lab head, service or just to make review of the operating and centering manuals for your system.

https://doi.org/10.1515/9783111317014-008

8.3 After tilting the sample, the image disappeared

If the sample exhibits significant magnetic features, it may deflect the condenser beam in different way, depending on the sample tilt. When working with magnetic material (and in fact even with most ferromagnetic steels), after setting zero focus, you should immediately verify and correct the position of the rotation center (Section 6.8). If you want to conduct diffraction studies using tilt, it is worth making sure that we are constantly working on only one side of the tilt (either from zero toward negative angles, or from zero toward positive angles). The greatest beam deflection occurs when sample passes the zero point and you should try to avoid this situation as much as possible. Of course, there is nothing stopping you from performing such tilt, but slowly and with the awareness of potential difficulties in controlling the position of the beam.

8.4 The image is defintely less sharp than usual

First, you should verify the correctness of the image focus and astigmatism (Section 6.11), after making sure that the condenser beam is correctly centered (Section 6.9). In the next step, it is worth making sure that our objective aperture is also centered (Section 6.10). If its edge is too close to the central beam in the diffraction pattern, astigmatism can be expected to be difficult to correct. The astigmatism setting could be also significantly different for magnetic and nonmagnetic samples.

   If the actions above have not been effective, it is worth installing a known, proven, and well-tested sample to confirm that the problem also occurs there. If this is the case and our microscope is equipped with a thermal emission gun, it is worth considering setting the appropriate emission current, voltage on the Wenhelt cylinder (a parameter usually called "bias") and the centering of gun shift and gun tilt. This type of procedure should be described in your microscope's operation manual and/or known to more experienced users. As a rule, insufficient work parameters of thermal gun may cause a doubled or tripled, superimosed images image, which manifests itself as reduced sharpness or is sometimes confused with excessive astigmatism.

8.5 The shape of the focused beam is unsatisfactory

If the focused beam is not a round spot but an ellipse or even shows several focal points of incorrect shape, the condenser astigmatism should first be verified (Section 6.4). If the astigmatism is correct, or its setting even intensifies the incorrect beam shape, the operating parameters of the thermal emission gun should be addressed (Section 8.4).

   If the focused beam vibrates and pulsates, and this situation has not happened before, for the LaB6 thermal emitter it may mean that it is time to replace it. For thermal emission guns, this situation may also be caused by excessive contamination of the electron gun (most likely in microscopes with oil diffusion pumps) and may require cleaning. I recommend contacting the service or more qualified users. It is worth remembering that in the case of thermal emission gun, replacing the cathode and cleaning the area around the gun usually does not require service intervention and is a normal maintenance activity, described in the user manual. It requires some caution and technical knowledge, but it should not pose any difficulties for a prepared and attentive user.

8.6 There is a problem with the calibration of the diffraction pattern

If you have a significant problem with the description and interpretation of electron diffraction, it is worth performing diffraction in a short period of time from the tested place of the sample and from the callibration sample respectively (usually a thin layer of polycrystalline aluminum or gold sputtered onto a carbon film) to set the defocus and other imaging parameters in a similar manner. Then it is easy to compare the interplanar distances of the standard and the sample, and minimize the risk of human error. If you really want to precisely calibrate single-crystal diffraction, you can sputter a few nanometers of polycrystalline gold on the sample you are interested in in a vacuum sputter (also used, for example, in the SEM sample preparation). You can even sputter minimally thin layers several times until the polycrystalline rings are visually visible in the diffraction mode. The additional layer will significantly complicate BF/DF observations and will constitute an important artifact of BF/DF images, but on SAEDs we will be dealing with both sample diffraction spots and the rings of the calibration sample at the same time. This allows the TEM diffraction error to be reduced to the lowest value possible. Of course, the sputtered material should be selected so that its reflections do not overlap too much with the reflections of the phase being tested.

9 I operate a transmission electron microscope – what next?

If you have reached this part of the book, you are probably well prepared for your first attempts at working on TEM. I encourage you to further deepen your knowledge, but I would like to emphasize that nothing influences the quality of research data more than long hours, days, and months of independent work with TEM. During the first attempts at TEM operation, it takes some time to feel relatively comfortable, and after some time you will gain the ability to freely solve simple issues (which appear sooner or later). If you only need the ability for simple nanomaterials characterization, you will probably quickly be satisfied with the progress and results achieved. If you plan to work on multiphase, metallic thin foils, or more demanding materials, it is the start of a long but interesting adventure. At the end of this chapter, I will once again list literature worth reading (led by the truly magnificent book [1] by David B. Williams and C. Barry Carter).

I also encourage you to try working on a variety of devices, both modern and slightly older ones. It is also worthwhile to take the opportunity to test new solutions at events and seminars organized by manufacturers of electron microscopes and side accessories like specimen holders. Finally, participation in leading electron microscopy conferences is a very valuable experience. Apart from the largest Microscopy&Microanalysis (organized by the Microscopy Society of America), Microscopy Conference (organized jointly by the German Society for Electron Microscopy, Swiss Society for Optics and Microscopy and Austrian Society for Electron Microscopy) and Electron Microscopy Congress (organized by the European Microscopy Society), it is worth attend one of the conferences organized by your local organization of researchers interested in electron microscopy techniques. These types of meetings are exceptionally conducive to networking, searching for new inspirations and ideas, as well for seeing how researchers who are more experienced than us work on their research problems. A beginner may feel intimidated by the abundance of modern techniques and advanced analytical methods, but it is worth remembering that even the simplest observation methods (even SAED/BF/DF!) can often lead to very valuable research.

If you plan to further develop yourself as a researcher in the field of electron microscopy, consider further education in scanning transmission electron microscopy (STEM) and energy dispersive spectroscopy (EDS, EDX) [2] or electron energy loss spectrometry (EELS) techniques. It wouldn't hurt to gain more experience, for example in using SEM, which can be useful even for checking the surface of TEM samples after preparation, or for supporting EDS testing if your TEM is not equipped with it. It may also be a good idea to review knowledge about sample preparation [3, 4] and electron diffraction [2, 5, 6].

https://doi.org/10.1515/9783111317014-009

I encourage exceptionally conscientious and interested microscopists to continue their education in accordance with their needs and profile. Scientists working more in the field of life sciences may be interested in immunolocalisation [7], 3D imaging techniques using tomography (easier for biological samples than for material science) [8] or cryogenic microscopy (cryoEM) [9, 10]. In the case of the latter technique, it is widely used mainly in the imaging and characterization of biomacromolecules using the Single Particle Analysis (SPA) protocol, but it is also used, among others, for imaging soft, bio-inspired nanomaterials. In turn, researchers with a strong background in materials science, physics, and chemistry may become more interested in in situ TEM [11], in which the entire, complex and often multistage experiment is carried out inside the microscope, directly during observation. This often requires constructing a very complex microscope configuration from scratch or purchasing expensive, dedicated specimen holders [12, 13]. A completely separate topic may also be the issue of advanced imaging of magnetic materials (using Lorentz microscopy, electron holography or differential phase contrast methods), or methods of effective observation at the highest possible magnifications, often requiring equipping the TEM with spherical aberration correctors [1]. These are just a few examples of specialized and demanding electron microscopy methods, each of which definitely deserves a separate, comprehensive study. I encourage you to find your own unique path and rely on the best resources available, some of which I list below. I hope this is the beginning of a beautiful, demanding, but satisfying adventure with transmission (and not only) electron microscopy techniques.

Literature

[1] Williams DB, Carter CB. Transmission electron microscopy: A textbook for materials science. 2nd ed. New York: Springer; 2008.

[2] Fultz B, Howe JM. Transmission electron microscopy and diffractometry of materials. 3rd ed. Berlin, New York: Springer; 2008. (Advanced texts in physics, 1439–2674).

[3] Ayache J, Beaunier L, Boumendil J, Ehret G, Laub D. Sample preparation handbook for transmission electron microscopy techniques. New York, NY: Springer New York; 2010.

[4] Ayache J, Beaunier L, Boumendil J, Ehret G, Laub D. Sample preparation handbook for transmission electron microscopy methodology. New York, NY: Springer New York; 2010.

[5] Andrews KW, Dyson, D.J., Keown, S.R. Interpretation of electron diffraction patterns. Springer Science+Business Media, LLC; 1967.

[6] Hirsch PB. Electron microscopy of thin crystals. Robert E. Krieger, Malabar, Florida; 1977.

[7] Schwartzbach SD, Osafune T. Immunoelectron microscopy. Totowa, NJ: Humana Press; 2010. (vol 657).

[8] Frank J, editor. Electron tomography: Methods for three-dimensional visualization of structures in the cell. 2nd ed. New York, NY: Springer; 2006.

[9] Dykstra MJ, Reuss LE. Biological electron microscopy: Theory, techniques, and troubleshooting. Springer Science+Business Media, LLC; 2003.

[10] Kuo J. Electron microscopy: Methods and protocols. 3rd ed. Totowa, NJ: Humana Press; 2014. (Methods in molecular biology; vol 1117).

[11] Ross FM. In Situ Transmission Electron Microscopy. In: Science of microscopy. Springer, New York, NY; 2007. p. 445–534.

[12] Ross FM. Liquid cell electron microscopy. Cambridge University Press; 2016.

[13] Hansen TW, Wagner JB. Controlled atmosphere transmission electron microscopy. Cham: Springer International Publishing; 2016.

Index

aberration
– chromatic 14–17, 25
– spherical 14, 23, 58, 82
accelerating voltage 6–8, 10, 13–14, 44, 58
anticontamination device 18, 25, 43, 63
aperture
– diffraction 24, 44, 66–68, 72, 74–76
– objective 13, 19, 24, 48, 53–56, 59, 62, 65, 68–74,
    76, 78
astigmatism 19, 23–24, 28, 42, 47–48, 56, 59–61,
    63, 70, 78
– condenser 23, 47, 53, 59, 78
– diffraction 70
– image 56

beam
– damage 12
bias 6, 78
bright field 24, 44, 52–53, 55, 65, 67, 70–75, 79, 81

charging 24
contamination 17–19, 24, 28, 36, 43–44, 49–50, 56,
    61, 79
contrast
– scattering 24, 54

dark field 24, 42, 56, 65, 70–76, 79, 81
diffraction 5, 13, 20–21, 23–24, 27–28, 42, 44, 49,
    52–56, 65–76, 78–79, 81

EDS 19–20, 48, 62, 81
electron beam 6–8, 10–11, 19–21, 27–31, 37, 42–44,
    58, 71, 76–77
– convergent 5
– intensity 22–24, 45–46, 53–54, 61–62, 69, 71, 77
– tilt 42, 52, 71, 73–74, 76
electron camera 11, 20–21, 24, 43, 60, 68
electron gun
– cold field 14–15, 17–18, 25, 44
– field emission 14, 17–18, 43–44, 48
– LaB6 13–14, 16–18, 25, 43–44, 78
– Schottky 14–17, 25, 44
– thermonic 48
– tungsten 6–7, 12, 14–18, 43, 49, 62
– Wenhelt cylinder 6, 78
electron lens 6, 10, 12

– condenser 5–6, 9–10, 21–25, 41–42,
    44–48, 53–54, 59, 61, 67–69, 71–73,
    77–78
– objective 9, 23, 28, 41–42, 44–45, 51, 54–55, 58,
    62, 76
– projector 5, 9–10, 21, 23–24, 41–42, 53

fluorescent screen 10–11, 14, 20–21, 27, 43, 45, 51,
    55, 58, 68, 73
focus 10, 19, 22–24, 41–42, 44–48, 50–54, 56–61,
    63, 67–72, 77–78
– diffraction 69
– image 23, 56, 58, 60, 69, 72
– in-focus 23–24, 56, 58, 60, 67,
    69, 72
– overfocus 24, 47, 56–59
– Scherzer 23, 54, 58
– underfocus 23–24, 46, 56–59
focused ion beam 28–29, 37–38

grid 29, 31, 39, 44–45, 77
– carbon 23–24, 29, 39, 56–57, 61, 79
– Formvar 30
– SiN 29, 31–32, 49
grinding
– dimple 34–37, 39
– mechanical 33

hole 6, 10, 24–25, 30–31, 36, 45, 49–50, 53, 55, 57,
    59, 77

image aquisition 10, 21, 59–61, 69
in situ 18, 20, 31, 38, 82

magnification 5, 9–10, 13, 21–23, 29, 44–45, 50,
    53–54, 56, 62–63, 68, 72, 77
maintenance 10, 17–18, 43, 47, 51, 63, 70,
    77, 79
microscope
– electron 5, 7–13, 81
– light 5, 9, 11–13, 28, 49

polishing 33–36, 39
– electropolishing 28, 35, 39
– ion 28, 36–37, 39
– mechanical 34

https://doi.org/10.1515/9783111317014-010

www.ingramcontent.com/pod-product-compliance
Lightning Source LLC
Chambersburg PA
CBHW081552220326
41598CB00036B/6649